Velázquez Spanish and English Glossary for the Science Classroom

Español – Inglés/English – Spanish

Velázquez Spanish and English Glossary for the Science Classroom

Español – Inglés/English – Spanish

Velázquez Press:
9682 Telstar Ave. Ste 110
El Monte, CA 91731 USA

Visit www.VelazquezPress.com or www.AskVelazquez.com

ISBN 10: 1-59495-010-5
ISBN 13: 978-1-59495-010-0

Printed in the United States of America

Third Edition

19 18 17 16 15 14 3 4 5 6

Library of Congress Control Number: 2011928093

List of Contributors

Lexicon researched and compiled by
Arthur C, Rachel L and the editors at
Velázquez Press

Thanks to Ruth García-Lago for her review of
Spanish-English terms

Thanks to Diego A Torres for the cover design

Preface

Limited English speakers in United States schools are faced with a double hurdle. They not only need to learn a second language in an ESL setting, but also must keep up with their peers in terms of content. In the case of abstract thought, such as science, a student may enter an ESL program already with well-formed concepts in his or her native language. The student and the teacher must bridge this gap, especially if this student is to succeed in the current American educational system.

The *Velázquez Spanish and English Glossary for the Science Classroom* is a newly revised work with a breadth of vocabulary that all bilingual students and teachers will find indispensable as a supplementary resource. This new bilingual glossary is convenient and relevant to classroom study and examinations because it offers a comprehensive array of lexical terms and translations. Here, the ESL student and science teacher will find not only technical terminology presented in English and Spanish, but also entries for the type of discourse most commonly used to discuss science.

This glossary fills a specific niche in elementary, middle and high school ESL and bilingual education by aspiring to the following objectives. First, this glossary is a resource used by limited English speaking students in the classroom to aid them in the comprehension of science terms and in their own expression of the language of science. Second, it is designed to be used by these same students during standardized science exams. Third, schools and districts provide this resource to parents as a means to increase parent involvement. Parents can use the glossary to support student homework assignments. Finally, the *Velázquez Spanish and English Glossary for the Science Classroom* can be used as a handy bibliographic reference for bilingual science teachers who must overcome questions of translation equivalents of specific terminology. Look no further than the *Velázquez Spanish and English Glossary for the Science Classroom* for all student, teacher and administrative science needs!

Velázquez Press

Prefacio

Los estudiantes que asisten a escuelas norteamericanas y que hablan un inglés limitado tienen que confrontar principalmente dos obstáculos: no sólo deben aprender el inglés como una segunda lengua sino que también deben mantenerse al nivel de sus compañeros de clase en las demás materias. En los casos de asignaturas de pensamiento abstracto, como lo son las ciencias, un estudiante puede ingresar a cualquier programa bilingüe con ciertos conceptos bien formados en su primera lengua. El estudiante y el maestro deben trabajar para hacer más pequeña esta diferencia lingüística, especialmente si el estudiante quiere tener éxito dentro del sistema educativo estadounidense.

El glosario bilingüe de términos científicos para uso en salones de clase, *Velázquez Spanish and English Glossary for the Science Classroom*, es una compilación revisada que presenta un vocabulario amplio, indispensable como recurso suplementario para todo estudiante y maestro bilingüe. Este nuevo glosario bilingüe resulta muy conveniente y pertinente para su uso tanto en salones de clase como en exámenes estandarizados, debido a su gran variedad de términos y traducciones comprensible. En este glosario, el estudiante de inglés como segunda lengua y su maestro de ciencias no sólo encontrarán terminología técnica presentada en inglés y español sino también distintas palabras sobre el tipo de discurso que se emplea comúnmente al hablar sobre las ciencias.

Este glosario cubre una necesidad especial de las escuelas primarias, secundarias y preparatorias que cuentan con educación bilingüe y aspira a los siguientes objetivos. Primero, es un recurso que los estudiantes con inglés limitado usarán en su salón de clase para ayudarles a comprender términos científicos, abstracciones y a expresarse en el lenguaje técnico correctamente. Segundo, está diseñado para que estos mismos estudiantes puedan utilizarlo durante los exámenes estandarizados de ciencias. Tercero, escuelas y distritos lo proveen a padres como herramienta para incrementar su participación en la educación de estudiantes. Los padres lo utilizan para traducir y entender vocabulario en inglés, y para ayudar al estudiante con su tarea. Por último, los maestros bilingües de ciencias que quieran aclarar dudas sobre traducciones de terminología específica tendrán a la mano como recurso bibliográfico el *Velázquez Spanish and English Glossary for the Science Classroom.* ¡No busque más! El *Velázquez Spanish and English Glossary for the Science Classroom* resolverá toda duda técnica como estudiante, maestro o administrador bilingüe.

Velázquez Press

User's Guide

This glossary is unique in that its word-to-word translation format meets requirements for school standardized testing. Disallowed are pronunciation keys, parts of speech, and guide words.

This glossary includes bold face entry words in alphabetical order. Homonyms that are different parts of speech are denoted with superscript. Commas are used to list synonymous translations. Related terms are indented beneath the main entry word in bold face to facilitate searchability.

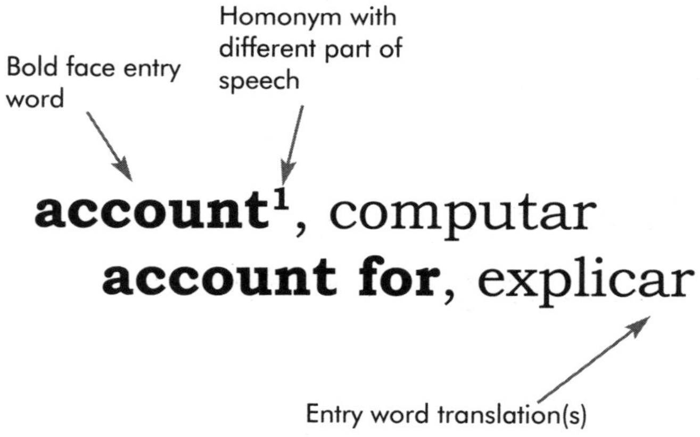

Bold face entry word

Homonym with different part of speech

account[1], computar

  account for, explicar

Entry word translation(s)

Guía del usuario

Este glosario es único ya que su formato de presentar las traducciones palabra por palabra cumple los requisitos para exámenes escolares estandarizados. No están permitidas la pronunciación fonética, las categorías gramaticales ni las palabras para guiar la traducción.

Este glosario incluye vocablos en orden alfabético que están en negrita. Los homónimos que tienen diferentes categorías gramaticales están indicados con un superíndice. Las comas se usan para enlistar los equivalentes sinónimos. Las palabras asociadas están en sangría debajo de la entrada principal y en negrita para facilitar la búsqueda.

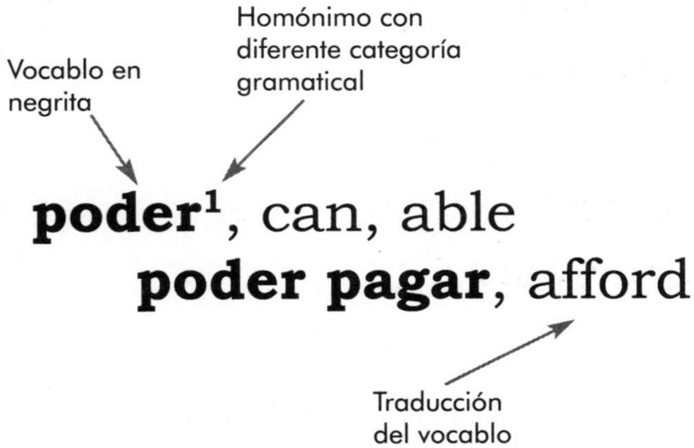

Vocablo en negrita

Homónimo con diferente categoría gramatical

poder[1], can, able

  poder pagar, afford

Traducción del vocablo

Section I
Español – Inglés

ábaco, abacus
abandonar, abandon
abanico aluvial, alluvial fan
abarcar, span
abarrancamiento, badlands
abastecer, stock
abdomen, abdomen
abedul, birch
abeja, bee
 abeja obrera, worker bee
 abeja reina, queen bee
abejón, drone
aberración, aberration
 aberración cromática, chromatic aberration
 aberración esférica, spherical aberration
abierto, open
abiogénesis, abiogenesis
ablación, ablation
ablandamiento del agua, water softening
abomaso, abomasum
abono, compost
aborto, abortion
abrasión, abrasion
abrazadera, clamp
abreviatura, abbreviation
absceso, abscess
abscisa, abscissa
absoluto, absolute
absorber, absorb
absorción, absorption
acampar, camp
acanalar, flute
acantilado, sea cliff
acanto, silver-thistle
ácaro, mite
acarrear, haul
acarreos fluvio-glaciáricos, outwash
acceso, access
accidente geográfico, landform
acción, action
 acción capilar, capillary action
 acción de masas, mass action
 acción de reflejo simple, simple reflex action
 acción glaciar, glacial action
 acción voluntaria, voluntary action
acechar, stalk
aceite, oil
 aceite de silicona, silicone oil
 aceite lubricante, lubricating oil
 aceite mineral, mineral oil
aceleración, acceleration
 aceleración de la gravedad, acceleration of gravity
 aceleración uniforme, uniform acceleration
acelerador de partículas, particle accelerator
acelerador lineal, linear accelerator
acelerar, accelerate
acentuar, accentuate
aceptor de hidrógeno, hydrogen acceptor
aceptor de protones, proton acceptor
acera, sidewalk

acero, steel
acero cromado, chrome steel
acero inoxidable, stainless steel
acetaldehído, acetaldehyde
acetato, acetate
acetato de polivinilo, polyvinyl acetate
acetilcolina, acetylcholine
acetileno, acetylene
acetona, acetone
acidez, acidity
ácido, acid
ácido acético, acetic acid
ácido acetilsalicílico, acetylsalicylic acid
ácido acrílico, acrylic acid
ácido ascórbico, ascorbic acid
ácido benzoico, benzoic acid
ácido bórico, boric acid
ácido butírico, butyric acid
ácido carbólico, carbolic acid
ácido carbónico, carbonic acid
ácido carboxílico, carboxylic acid
ácido cianhídrico, hydrocyanic acid, hydrogen cyanide
ácido cianuro de hidrógeno, hydrogen cyanide
ácido cítrico, citric acid
ácido clorhídrico, hydrochloric acid
ácido crómico, chromic acid
ácido débil, weak acid
ácido desoxirribonucleico (ADN), deoxyribonucleic acid (DNA)
ácido dicarboxílico, dicarboxylic acid
ácido esteárico, stearic acid
ácido fólico, folic acid
ácido fórmico, formic acid
ácido fuerte, strong acid
ácido graso, fatty acid
ácido hipocloroso, hypochlorous acid
ácido láctico, lactic acid
ácido mineral, mineral acid
ácido nítrico, nitric acid
ácido nitroso, nitrous acid
ácido nucleico, nucleic acid
ácido oleico, oleic acid
ácido orgánico, organic acid
ácido oxálico, oxalic acid
ácido pirúvico, pyruvic acid
ácido polibásico, polybasic acid

ácido propiónico, propionic acid
ácido ribonucleico (ARN), ribonucleic acid (RNA)
ácido salicílico, salicylic acid
ácido sulfhídrico, hydrogen sulfide
ácido sulfúrico, sulfuric acid
ácido tartárico, tartaric acid
ácido tereftálico, terephthalic acid
ácido ternario, ternary acid
ácido úrico, uric acid
acné, acne
acompañar, accompany
acondicionamiento, conditioning
acoso sexual, sexual harassment
acotamiento, shoulder
acromaticidad, achromaticity
acromegalia, acromegaly
actínido, actinide
actinio, actinium
actitud, attitude
actividad, activity
actividad biológica, biological activity
actividad de la superficie, surface activity
actividades humanas, human activities
acto motor, motor action
acto reflejo del estornudo, sneezing reflex
(el) acto sexual, sexual intercourse
actualmente, currently
acuario, aquarium
acuático, aquatic
acuerdo, arrangement
acuicultura, aquaculture
acuífero, aquifer
acumular, accumulate
acupuntura, acupuncture
acústica, acoustics
adaptación, adaptation
adaptación asimilada, learned adaptation
adaptación estructural, structural adaptation
adaptación heredada, inherited adaptation
adaptar, adapt
adaptativo, adaptive
adecuado, adequate
además, furthermore
adenina, adenine
adenoides, adenoids
adenosín monofosfato, adenosine monophosphate (AMP)
adenosín trifosfato, adenosine triphosphate (ATP)
adhesivo, adhesive
adiabático, adiabatic
adicción, addiction
adicional, additional
adiposo, adipose
aditivo, additive
adjuntar, enclose
administrar, manage
admitir, allow
ADN, DNA
ADN polimerasas, DNA polymerase
adolescencia, adolescence
adquirir, acquire, gain
adrenalina, adrenaline

adsorción, adsorption
adulto, adult
advertir, note
adyacente, adjacent
aerobio, aerobic, aerobe
aerodinámico, aerodynamic
aeroespacio, aerospace
aeronáutica, aeronautics
aerosol, aerosol
afectar[1], affect
afectar[2], touch
afelio, aphelion
afloramiento, outcrop
aflorar, surface
afluente, tributary
agalla, gall
agar, agar
agarosa, agarose
ágata, agate
agente, agent
    agente antiespumante, antifoaming agent
    agente de desarrollo, developing agent
    agente de refuerzo, reinforcing agent
    agente de secuestro, sequestering agent
    agente deshidratante, dehydrating agent
    agente espumante, blowing agent
    agente extintor, extinguishing agent
    agente mutagénico, mutagenic agent
    agente tensioactivo, surface active agent
    agentes atmosféricos, weathering agents
aglutinógeno, agglutinogen
agnatos, jawless fish
agobiar, swamp
agotamiento, depletion
    agotamiento del suelo, soil depletion
agotar[1], deplete
agotar[2], exhaust
agrandar, magnify
agricultura de contorno, contour farming
agrietamiento, cracking
agrietar, crack
agrimonia[1], liverwort
agrimonia[2], silver-weed
agrio, sour
agrupamiento, grouping
agrupar, group
agua, water, aqua
    agua de cristalización, water of crystallization
    agua de deshielo, meltwater
    agua de hidratación, water of hydration
    agua del sistema vascular, water vascular system
    agua desionizada, deionized water
    agua dulce, freshwater
    agua dura, hard water
    agua mineral, mineral water
    agua pesada, heavy water
    agua salada, saltwater
    agua salobre, brackish water
    agua subterránea, groundwater
    agua superenfriada, supercooled water

aguas residuales, sewage
aguas termales, hot spring
aguanieve, sleet
agudo, high
aguijón, stinger
águila, eagle
aguilera, eyrie
aguja, needle
aguja de coser, sewing needle
aguja de pino, pine needle
aguja hipodérmica, hypodermic needle
agujero negro, black hole
ahumar, smoke
aire, air
aireación, aeration
airear, aerate
aislado, isolated
aislamiento, isolation
aislamiento geográfico, geographic isolation
aislamiento reproductivo, reproductive isolation
aislante, insulator
aislar, insulate, isolate
ajustado, tighten
ajustar, adjust
ajuste fino, fine adjustment
ajuste grueso, coarse adjustment
ala, wing
alambique, pipe still
alambre de gasa, wire gauze
álamo temblón, aspen
alantoides, allantosis
alargar, lengthen
alarma antirrobo, burglar alarm
alarmante, startling
albino, albino
albumen, albumen
albúmina, albumin
alcalescencia, alkalescence
álcali, alkali
alcalino, alkaline
alcaloide, alkaloid
alcance, range
alcanfor, camphor
alcano, alkane
alcano-derivados, alkane derivative
alcanzar, attain
alcohol, alcohol
alcohol amílico, amyl alcohol
alcohol bencílico, benzyl alcohol
alcohol butílico, butyl alcohol
alcohol de madera (metanol), wood alcohol
alcohol de polivinilo, polyvinyl alcohol
alcohol desnaturalizado, denatured alcohol
alcohol etílico, ethyl alcohol
alcohol graso, fatty alcohol
alcohol industrial, industrial alcohol
alcohol isopropílico, isopropyl alcohol
alcohol metílico, methyl alcohol
alcohol primario, primary alcohol
alcohol secundario, secondary alcohol

alcohol terciario, tertiary alcohol
alcohol trihidroxi, trihydroxy alcohol
aldehídos, aldehyde
aleación, alloy
aleación de acero, alloy steel
aleación de plata, silver alloy
aleación fusible, fusible alloy
aleatorio, random
alejarse, recede
aleno, allene
alérgeno, allergen
alergia, allergy
alérgico, allergic
aleta[1], fin
aleta[2], flipper
aleta[3], tail fin
aletargado, dormant
alfabetización científica, scientific literacy
alféizar, sill
alfombra, carpet
alfombrilla de ratón, mouse pad
alga, alga
alga marina, seaweed
algas verdes, green algae
álgebra, algebra
álgebra booleana, Boolean algebra
algebraico, algebraic
algodón pólvora, guncotton
algoritmo, algorithm
aligátor, alligator
alimentar, nourish
alimento, food
alimentos alergénicos, food allergen
almacenamiento, storage
almacenamiento del suelo, soil storage
almacenar, store
almeja, clam
almidón, starch
alotropía, allotropy
alótropo, allotrope
alqueno, alkene
alquilo, alkyl
alquimia, alchemy
alquino, alkyne
alquitrán, tar
alrededor, around
alrededores, surroundings
altavoz, loudspeaker
alteración, alteration
alteración cromosómica, chromosomal alteration
alterado, altered
alterar, alter
alternador, alternator
alternar, alternate
altímetro, altimeter
altiplanicie, plateau
altitud, altitude
altitud absoluta, absolute altitude
alto[1], alto
alto[2], high
altura, height
altura de la Estrella Polar, altitude of Polaris
alucinógeno, hallucinogen
alud, avalanche
alumbre, alum
alúmina, alumina
aluminio, aluminum
alunizaje, moon landing
aluvión[1], alluvium
aluvión[2], flood

alvéolo, alveolus
alvéolo pulmonar, air sac
amalgama, amalgam
amarillo, yellow
amarras, moorings
amatista, amethyst
ámbar, amber
ambiente[1], atmosphere
ambiente[2], environment
ambiente[3], ambient
ambrosía, ragweed
ameba, amoeba
amianto, asbestos
amiba, ameba
amígdala, tonsil
amigdalitis, strep throat, tonsilitis
amilasa, amylase, ptyalin
amilasa salival, salivary amylase
aminoácido, amino acid
aminoácido esencial, essential amino acid
amnesia, amnesia
amniocentesis, amniocentesis
amnios, amnion
amoníaco, ammonia
amoníaco líquido, ammonia liquor
amonificación, ammonification
amonio, ammonium
amortiguador[1], buffer
amortiguador[2], shock absorber
amperímetro[1], ammeter
amperímetro[2], amperometry
amperio, ampere
ampliar, extend, widen
amplificación, amplification
amplitud, amplitude
anabolismo, anabolism
anaconda, anaconda
anádromo, anadromous
anaerobio, anaerobic, anaerobe
anafase, anaphase
analgésico, analgesic
análisis, analysis
análisis cualitativo, qualitative analysis
análisis cuantitativo, quantitative analysis
análisis gravimétrico, gravimetric analysis
análisis inorgánico, inorganic analysis
análisis instrumental, instrumental analysis
análisis polarográfico, polarographic analysis
análisis volumétrico, volumetric analysis
analista, analyst
analizar, analyze
analogía, analogy
analógico, analog
análogo, analogous
anatomía, anatomy
ancestro, ancestor
ancho, width
anciano, ancient
andamio, scaffold
andesita, andesite
andrógeno, androgen
anemia, anemia
anemia falciforme, sickle cell anemia
anemómetro, anemometer
anémona de mar, sea anemone
aneroide, aneroid
anestesia, anesthesia
aneurisma, aneurysm

anfetamina, amphetamine
anfibio, amphibian
anfíbol, amphibole
anfitrópico, amphiprotic
anfoterismo, amphoterism
anfótero, amphoteric
angina de pecho, angina pectoris
angiosperma, angiosperm
ángstrom, angstrom
anguila, eel
ángulo, angle
ángulo adyacente, adjacent angle
ángulo agudo, acute angle
ángulo central, central angle
ángulo crítico, critical angle
ángulo de bonos, bond angle
ángulo de incidencia, angle of incidence, heading angles
ángulo de insolación, angle of insolation
ángulo de reflexión, angle of reflection
ángulo de refracción, angel of refraction
ángulo exterior, exterior angle
ángulo interior, interior angle
ángulo llano, straight angle
ángulo obtuso, obtuse angle
ángulo recto, right angle
ángulos alternos, alternate angles
ángulos complementarios, complementary angles
ángulos conjugados, conjugate angles
ángulos suplementarios, supplementary angles
ángulos verticales, vertical angles
anhídrido, anhydride
anhídrido nitroso, nitric anhydride
anhídrido básico, basic anhydride
anhidro, anhydrous
añil, indigo
anillo, ring
anillo bencénico, benzene ring
anillo de crecimiento, growth ring (biology)
anillo de soporte, ring stand
animal, animal
animales con pezuñas, hoofed animal
anión, anion
ano, anus
año, year
año luz, light-year
año solar, solar year
ánodo, anode
anomalía, anomaly
anorexia, anorexia
anquilostoma, hookworm
antagonista, antagonist
antena, antenna
antepecho de ventana, window-sill
antera, anther

anterior, former, anterior
antiácido, antacid
antibiótico, antibiotic
anticiclón, anticyclone
anticlinal, anticline
anticloro, antichlor
anticonceptivo, contraceptive
anticonceptivos, contraception
anticongelante, antifreeze
anticuerpo, antibody
antídoto, antidote
antielectrón, positron
antienzima, antienzyme
antifebrin, antifebrin
antígeno, antigen
antiguo, ancient
antihistamínico, antihistamine
antilogaritmo, antilogarithm
antimonio, antimony
antinodo, antinode
antioxidante, antioxidant
antipartícula, antiparticle
antirretorno, anti-return
antiséptico, antiseptic
antitoxina, antitoxin
antracita, anthracite
ántrax, anthrax
antropología, anthropology
anual, annual
anular, annular
aorta, aorta
aparato circulatorio, circulatory system
aparato digestivo, digestive system
aparato eléctrico, appliance
aparato polarográfico, polarographic apparatus
aparato reproductor, reproductive system
apareamiento, mating
aparente, apparent
aparición, emergence
apatito, apatite
apéndice, appendix (pl. appendices)
apendicitis, appendicitis
aplanar, flatten
aplicar¹, apply
aplicar², allow
apogeo, apogee
apoplejía, stroke
apoyar, support
apoyo, support
aprender, learn
aprendido, learned
aprendizaje, learning
apretado, tighten
aprobación, approval
apropiado, appropriate
aproximadamente, approximately
aproximado, approximate
aptitud, ability
apuro, rush
arácnido, arachnid
árbol, tree
  árbol de hoja caduca, deciduous tree
arbusto, shrub
archivo, file
arcilla, clay
arco¹, arch
  arco aórtico, aortic arch
arco², arc
  arco insular, volcanic arc
  arco iris, rainbow
  arco reflejo, reflex arc
arco³, bow

ardilla, chipmunk
ardor de acidez, heartburn
ardor de estómago, heartburn
ardor de pirosis, heartburn
área, area
área de la superficie, surface area
área sensorial, sensory area
arena, sand
arenas movedizas, quicksand
arenisca, sandstone
arista, arête
argón, argon
árido, arid
Aristóteles, Aristotle
aritmética, arithmetic
arma nuclear, nuclear weapon
armadillo, armadillo
armado de garras, clawed
armadura, armature
armar, assemble
armazón, framework
armónico, harmonic
armonioso, harmonious
ARN, RNA
ARN de transferencia (tARN), transfer RNA (tRNA)
ARN mensajero, messenger RNA (mRNA)
ARN ribosomal, ribosomal RNA
ARNm (ARN mensajero), mRNA (messenger RNA)
aromático, aromatic
arquea, archaeon
arqueología, archaeology
arquero, archer
arquitecto, architect
arrastrar, haul
arrastre, drag
arrecife, reef
arrecife de coral, coral reef
arreglo, array
arriba, upward
arrojar, cast
arruga, wrinkle
arsénico, arsenic, arsenical
arseniuro, arsenide
arseniuro de galio, gallium arsenide
arteria, artery
arteria carótida, carotid artery
arteria coronaria, coronary artery
arteria hepática, hepatic artery
arteria pulmonar, pulmonary artery
arteria renal, renal artery
arteriola, arteriole
arteriosclerosis, arteriosclerosis
articulación, joint
articulación de la cadera, hip joint
articulación de la rodilla, knee joint
articulación de rótula, ball and socket joint
articulación del codo, elbow joint
articulación del hombro, shoulder joint
artificial, artificial
artillería, artillery
artiodáctilo, artiodactyls
artritis, arthritis
artritis reumatoide, rheumatoid arthritis

artrópodo, arthropod
asa de Henle, loop of Henle
asbesto, asbestos
ascáride, roundworm
ascensión recta, right ascension
asegurar, ensure
asentar, settle
aséptico, aseptic
asesoramiento genético, genetic counseling
asexual, asexual
asexualmente, asexually
asfalto, asphalt
asfixia, asphyxia
asiento abatible, swing seat
asignar, allow
asimetría, asymmetry
asimilación, assimilation
asíntota, asymptote
asistente, assistant
asma, asthma
asociación, association
  asociación de suelos, soil association
asociar, associate
aspecto, aspect
  aspecto afectivo, bonding
aspiradora, vacuum cleaner
aspirar, aspirate
aspirina, aspirin
astenósfera, asthenosphere
aster, aster
asteroide, asteroid
astigmatismo, astigmatism
astrágalo, talus
astral, stellar
astringente, astringent
astroblema, impact crater
astrofísica, astrophysics
astrometría, astrometry
astronomía, astronomy
astronómico, astronomical
astrónomo, astronomer
asunto, case
ataque cardíaco, heart attack
atar, bind
atávico, atavistic
ataxia, ataxia
atleta, athlete
atmósfera, atmosphere
atmosférico, atmospheric
atolón, atoll
atómico, atomic
átomo, atom
atornillar, screw
atracción, attraction
  atracción gravitacional, gravitational attraction
  atracción interiónica, interionic attraction
atractivo, attractive
atraer, attract
atrincherado, entrenched
atrincherar, trench
atrofia, atrophy
audiencia[1], hearing
audiencia[2], audience
auditivo, auditory
auditoría de datos, data auditing
auditorio, auditorium
augita, augite
aumentar, increase, magnify, raise
aumento, increase
aurícula, atrium, auricle
aurículas, atria
aurora, aurora
  aurora boreal, northern lights, aurora borealis
ausencia, absence
autismo, autism
autoclave, autoclave

autofecundación, self-fertilization
autoinductancia, self-inductance
autoinflamación, spontaneous combustion
autoinmune, autoimmune
automático, automatic
automóvil, automobile
autopolinización, self-pollination
autopsia, autopsy
autosoma, autosome
autótrofo, autotrophic
auxina, auxin
avalancha, avalanche
  avalancha de tierra, landslide
avances realizados, progress made
avanzar, advance
ave, bird
  ave de corral, fowl
  ave rapaz, raptor
avestruz, ostrich
aviar, avian
aviario, aviary
avión, airplane, aircraft
  avión a reacción, jet plane
avispa, wasp
axila, axil
axioma, axiom
axón, axon
azeotrópica, azeotropic
azimut, azimuth
azúcar, sugar
  azúcar en la sangre, blood sugar
  azúcares simples (monosacáridos), simple sugar (monosaccharide)
azufre, sulfur
azul, blue
  azul de metileno, methylene blue
  azul de Prusia, Prussian Blue
  azul eléctrico, peacock blue
  azul de bromotimol, bromthymol blue
azurita, azurite

B

babuino, baboon
bacteria, bacterium
bacteriana, bacterial
bacterias, bacteria
  bacterias fijadoras de nitrógeno, nitrogen-fixing bacteria
  bacterias nitrificantes, nitrifying bacteria
  bacterias desnitrificantes, denitrifying bacteria
bacteriófago, bacteriophage
bahía, bay
bajar el cursor, scroll down
bajo, low
bala, bullet
  bala de cañon, cannonball
balance hídrico, water budget
balance térmico, heat budget
balance radiativo, radiative balance
balancear, balance

balancearse, balance, swing
balanceo, rocking
balanza, balance, scale
balanza analítica, analytical balance
balanza de doble plato, double-pan balance
balanza de torsión, torsion balance
balanza de tres brazos, triple beam balance
balanza electrónica, electronic balance
balística, ballistics
ballena, whale
bambú, bamboo
bancales, terraced
banco de arena[1], sandbar
banco de arena[2], shoal
bandas elásticas, banding
baño de agua caliente, hot water bath
baño de plata, silver plate
banquina, shoulder
barba de ballena, baleen
barba de ballena, whalebone
barbilla, chin
baricentro, barycenter
bario, barium
bariónico, baryon
barita, barite
baritina, barite
barómetro, barometer
barómetro aneroide, aneroid barometer
barómetro de mercurio, mercury barometer
barra de control, control rod
barra espaciadora, space bar (computer)
barranco, gully
barrer, sweep
barrera, barrier
barrera de coral, barrier reef
barrera del sonido, sound barrier
barrera sónica, sonic barrier
barril, barrel
basalto, basalt
báscula mecánica, beam balance
base, base, basis
base débil, weak base
base sólida, strong base
básico, basic
bastón (célula), rod
basura, garbage
basurero, dump
batería, battery
batiscafo, bathyscaphe
batolito, batholith
bauxita, bauxite
baya, berry
bayou, bayou
bazo, spleen
becquerel, becquerel
benceno, benzene
bencidina, benzidine
beneficio, benefit
beneficioso, beneficial
benigno, benign
bentos, benthos
benzaldehído, benzaldehyde
benzenoide, benzenoid
beriberi, beriberi
berilio, beryllium
berilo, beryl
berkelio, berkelium
berma, berm
beta-caroteno, beta carotene
betún, bitumen
bibliotecario, librarian

bicapa lipídica, lipid bilayer
bicarbonato de sodio, baking soda
bicarbonato sódico, sodium bicarbonate
bíceps, biceps
bicho, bug
bidón, drum
(el) bien común, common good
(el) Big Bang, Big Bang
bilis, bile
billar, billiard
billón, trillion
bimetálico, bimetallic
binaria eclipsante, eclipsing binary
binario, binary
binocular, binocular
binomio, binomial
biodegradable, biodegradable
biodiversidad, biodiversity
bioética, bioethics
biofísica, biophysics
biogénesis, biogenesis
biogeografía, biogeography
biología, biology
bioma marino, marine biome
biomas terrestres, terrestrial biome
biomasa, biomass
biomecánica, biomechanics
biónica, bionics
biopsia, biopsy
bioquímica, biochemistry
bioquímico, biochemical
biorremediación, bioremediation
biorritmo, biorhythm
biósfera, biosphere
biosíntesis, anabolism
biotecnología, biotechnology
biotecnológicas, biotechnological
biótico, biotic
biotina, biotin
biotita, biotite
bípedo[1], biped
bípedo[2], bipedal
bismuto, bismuth
bit, bit
bitácora, blog
bivalente, bivalent
bivalvo, bivalve
blanco, white
blanqueador, bleach
blástula, blastula
blog, blog
bloque, block, brick
bloquear, block
bloqueo numérico, number lock (computer)
bobina, coil
  bobina de inducción, induction coil
  bobina de Ruhmkorff, induction coil
  bobina primaria, primary coil
  bobina secundaria, secondary coil
boca, mouth
bocio, goiter
bola de boliche, bowling ball
bolita, pith ball
bolsa de aire, air pocket
bomba[1], bomb
  bomba atómica, atomic bomb
  bomba de hidrógeno, hydrogen bomb
bomba[2], pump
  bomba de calor, heat pump
  bomba hidráulica, hydraulic pump

bombardeo nuclear, nuclear bombardment
boquilla, mouthpiece
bórax, borax
borde convergente, convergent boundary
boreal, boreal
boro, boron
boroflouride, boroflouride
borosilicato, borosilicate
borrasca, trough, low pressure area
bosque, forest, wood
 bosque de algas marinas, kelp forest
 bosque templado caducifolio, temperate deciduous forest
 bosque de coníferas, coniferous forest
bosquejo, sketch
botánica, botany
botella, bottle
 botella de pesaje, weighing bottle
bovino, bovine
boyante, buoyant
branca ursina, silver-thistle
branquia, gill
braquiópodo, brachiopod
braza, fathom
brazo, arm
brazo muerto (de un río), oxbow
brea, tar
brecha, breccia, gap
breve, brief
brevemente, briefly
brillo, luster
 brillo perlado, pearly luster
brillómetro, gloss meter
brincar, spring
brisa de tierra, land breeze
brisa marina, sea breezes
brizna, blade
brócoli, broccoli
bromeliácea, bromeliad
bromo, bromine
bromuro, bromide
 bromuro de etileno, ethylene bromide
bronce, bronze
bronquio, bronchus
bronquiolo, bronchiole
bronquios, bronchi
bronquitis, bronchitis
brotar, spring
brote[1]**,** sprout
brote[2]**,** bud
brújula, compass
 brújula magnética, magnetic compass
bubón, bubo
bucear, dive
buche, crop
búho, owl
bulbo, bulb
 bulbo olfatorio, olfactory bulb
 bulbo raquídeo, medulla oblongata
bulimia, bulimia
bulto, bulge
buque, liner
burbuja, bubble
bureta, burette
búsqueda, search
butano, butane
butanodiol, butanediol
butanol, butanol
buteno, butene
butílico, butyl
buzo, diver
bypass, bypass
byte, byte

caballerizo, stable

caballo, horse

caballo de vapor, horsepower

caballo marino, sea horse

cabeza, head

cable, cable (computer)

cabo, cape, headland

cabra, goat

cachorro, puppy

cactus, cactus

cadena, chain

cadena abierta, open chain

cadena alimenticia, food chain

cadena lineal compuesta, straight-chain compound

cadena montañosa, mountain range

cadena ramificada, branch-chain

cadena respiratoria, respiratory chain

cadera, hip

cadmio, cadmium

caducifolio, deciduous

cafeína, caffeine

caída libre, free fall

caimán, alligator

caja de cartón, carton

caja de voz (laringe), voice box

caja torácica, rib cage

cajón, crate

calamar, squid

calcificación, calcification

calcinar, calcine

calcio, calcium

calcita, calcite

calcopirita, chalcopyrite

calculadora, calculator

calcular[1], estimate

calcular[2], calculate

cálculo[1], calculus

cálculo[2], estimation

cálculos biliares, gallstone

caldera, kettle

calentador de inmersión, immersion heater

calentamiento global, global warming

calentar, heat

calibrar, calibrate

calibre[1], caliper

calibre[2], gauge

calidad, quality

caliente, hot

californio, californium

callo, callus

calomel, calomel

calor, heat

calor de combustión, heat of combustion

calor de condensación, heat of condensation

calor de cristalización, heat of crystallization

calor de dilución, heat of dilution

calor de formación, heat of formation

calor de fusión, heat of fusion

calor de hidratación, heat of hydration

calor de reacción, heat of reaction

calor de sublimación, heat of sublimation

calor de transición, heat of transition
calor de vaporización, heat of vaporization
calor específico, specific heat
calor geotérmico, geothermal heat
calor latente, latent heat
caloría, calorie
calórico, caloric
calorimetría, calorimetry
calorímetro, calorimeter
celular, cell
cama, bed, bedding
camaleón, chameleon
cámara, camera, chamber
cámara de burbujas, bubble chamber
cámara de niebla, cloud chamber
cambiar[1]**,** change, alter
cambiar[2]**,** switch
cambio[1]**,** change
cambio de dirección, change of direction
cambio de longitud de una sombra, changing length of a shadow
cambio de movimiento, change of motion
cambio de sexo, sex change
cambio de temperatura adiabático, adiabatic temperature change
cambio de velocidad, change of speed
cambio físico, physical change
cambio químico, chemical change
cambio químico espontáneo, spontaneous chemical change
cambios cíclicos, cyclic changes
cambios en el medio ambiente, environmental changes
cambio[2]**,** switch
cámbium, cambium
cámbium vascular, vascular cambium
Cámbrico, Cambrian
caminante, hiker
caminar por afloramientos, walking the outcrop
camión cisterna, tanker
camisa, shirt
campana de Gauss, bell curve
campista, camper
campo, field
campo del microscopio, field of microscope
campo eléctrico, electric field
campo escalar, scalar field
campo gravitacional, gravitational field
campo magnético, magnetic field
camuflaje, camouflage
caña, reed
canal, channel
canal auditivo, ear canal
canal de conducción, raceway, racetrack
canal de parto, birth canal

canal Haversiano, Haversian canal
canal radial, radiating canal
canal semicircular, semicircular canal
cáncer, cancer
cáncer de piel, skin cancer
cancerígeno, carcinogen
candela, candela
cangrejo de río, crayfish
canguro, kangaroo
canino, canine
canoa, canoe
cañón[1], canyon
cañones submarinos, submarine canyons
cañón[2], gorge
cantidad de movimiento, momentum
cantidad[1], amount
cantidad[2], quantity
cantidad vectorial, vector quantity
canto rodado, boulder
caolín, kaolin
caos, chaos
capa[1], layer
capa de arbusto, shrub layer
capa de hielo, ice sheet
capa de ozono, ozone layer
capa de cera, wax layer
capa en empalizada, Palisade layer
capa freática, water table
capa esponjosa, spongy layer
capas, layering
capa[2], mantle
capa[3], shell
capacidad, capacity, ability
capacidad vital, vital capacity
capacitación, capacitance
caparazón, carapace
capaz, capable
capilar, capillary
capilaridad, capillarity, capillary action
cápsula, capsule
cápsula de Bowman, Bowman's capsule
captar, capture
captura, capture
capullo[1], bud
capullo[2], cocoon
caracola, conch
característica, characteristic, feature, trait
caracteristicas adquiridas, acquired characteristics
características del paisaje, landscape features
características sexuales secundarias, secondary sexual characteristics
carbeno, carbene
carbocíclico, carbocyclic
carbohidrasa, carbohydrase
carbohidrato, carbohydrate
carbohidrato complejo, complex carbohydrate
carbón, coal
carbón vegetal, charcoal
carbonatación, carbonation
carbonato, carbonate

carbonato de bario, barium carbonate
carbonato de calcio, calcium carbonate
carbonato de carbono, carbonate carbonato
carbonato de sodio, soda ash
carbonato sódico, sodium carbonate
carbonido, carbonide
carbonífero, carboniferous
carbono (C), carbon (C)
carboxilo, carboxyl
carbunco, anthrax
carbunclo[1], carbuncle
carbunclo[2], anthrax
carburación, carburetion
carburo, carbide
carburo de boro, boron carbide
carburo de tungsteno, tungsten carbide
carburo de calcio, calcium carbide
carburo de silicio, silicon carbide
carcinógeno, carcinogen
cardíaco, cardiac
cárdidos, cockle
cardiología, cardiology
carecer de, lack
carga[1], charge
carga estática, static charge
carga negativa, negative charge
carga por inducción, charging by induction
carga positiva, positive charge
carga por conducción, charging by conduction
carga[2], load, strain
carga de fondo, bedload
cargado, charged
caries, tooth decay
cariotipo, karyotyping
carnívoro[1], carnivorous
carnívoro[2], carnivore
caroteno, carotene
carpelo, carpel
carpeta, binder
carpo, carpal
carril de aire, air track
carroñero, scavenger
Carso, Carso
cartílago, cartilage
cascada, waterfall
cáscara, shell
caso, case
casos declarados, reported cases
casos importados, imported cases
casta, caste
castor, beaver
catalasa, catalase
catalizador, catalyst
catalizadores biológicos, biological catalysts
catarata, cataract
catarina, ladybug
categoría, category
catéter, catheter
catión, cation
cátodo, cathode
caucho butílico, butyl rubber
caudal, stream flow
caudatum, caudatum
cauri, cowire
causa, cause
cavidad, cavity
cavidad gastrovascular,

gastrovascular cavity
cavidad nasal, nasal cavity
cavidad pleural, pleural cavity
cavidad sanguínea, blood cavity
cavitación, cavitation
caza, hunting
caza furtiva, poaching
CD-ROM, CD-ROM
cefálico, cephalic
cefalocordado[1], cephalochordate
cefalocordado[2], lancelet
cefalópodo, cephalopod
cefalotórax, cephalothorax
ceguera nocturna, night blindness
ceja, eyebrow
celda, cell
celda de combustible, fuel cell
celda eléctrica, electric cell
celda electrolítica, electrolytic cell
celda electroquímica, electrochemical cell
celda fotovoltaica, photovoltaic cell
celda galvánica, voltaic cell, galvanic cell
celda solar, solar cell
celentéreos, cnidarian
celeste, celestial
celosía, lattice
celtium, celtium
célula, cell
célula eucariota, eukaryotic cell
célula falciforme, sickle cell
célula germen, germ cell
célula hija, daughter cell
célula madre, parent cell, stem cell
célula melanocita, melanocyte cell
célula oclusiva, guard cell
célula olfativa, olfactory cell
célula plasmática, plasma cell
célula plastocito, plasma cell
célula T, T cell
célula urticante, stinging cell
células de Schwann, Schwann's cell
células madre embrionarias, embryonic stem cells
células somáticas, somatic cell
celulitis, cellulite
celuloide, celluloid
celulosa, cellulose
cementación, cementation
cementar[1], cement
cemento[2], cement
cénit, zenith
ceniza, ash
ceniza de soda, soda ash
ceniza volcánica, volcanic ash
centelleo, scintillation
centi-, centi-

centígrado, centigrade
centigramo, centigram
centilitro, centiliter
centímetro (cm), centimeter (cm)
  centímetro cúbico, cubic centimeter
central nuclear, nuclear power station
centrífuga, centrifuge
centrifugadora, centrifuge
centrífugos, motor nerve
centriolo, centriole
centrípeto, centripetal
centro, core, center
  centro de gravedad, center of gravity
  centro de inhibición, inhibition center
  centro de masas, center of mass
  centro de reflejo, reflex center
  centro neurálgico, nerve center
centrómero, centromere
cepa resistente, resistant strain
cera, wax
cercano, nearby
cerda, bristle
cerdo, pig
cereal, cereal
cerebelo, cerebellum
cerebral, cerebral
cerebro, brain
  cerebro anterior, forebrain
  cerebro medio, midbrain
cerio, cerium
cero, zero
  cero absoluto, absolute zero
cerro, hill
certero, certain
cervical, cervical
cérvix uterino, cervix
cesio, cesium
césped[1], turf
césped[2], grass
césped[3], lawn
cetáceo, cetacean
cetona, ketone
charco, puddle
chasquear, snap
chimenea[1], chimney
chimenea[2], smokestack
chimpancé, chimpanzee
chip, chip (computer)
  chip de silicio, silicon chip
chispa, spark
chiva, goat
chocar, collide
choque, shock
  choque elástico, elastic collision
chorro, jet
cian, cyan
cianhídrico, hydrocyanic
cianógeno, cyanogen
cianuro, cyanide
cibernética, cybernetics
cicádido, cicada
cíclico, cyclic
ciclismo, cycling
ciclo, cycle
  cíclo cardíaco, heartbeat cycle
  ciclo celular, cell cycle
  ciclo de las rocas, rock cycle
  ciclo de oxígeno, oxygen cycle

ciclo de vida, life cycle
ciclo de vida de una estrella, star life cycle
ciclo del agua, water cycle
ciclo del carbono, carbon cycle
ciclo del oxigeno-dióxido de carbono, oxygen-carbon dioxide cycle
ciclo diario, daily cycle
cíclo frecuencia cardíaca, heartbeat cycle
ciclo menstrual, menstrual cycle
cicloalcano, cycloalkane
ciclohexano, cyclohexane
ciclón, cyclone
ciclones de latitudes medias, mid-latitude cyclone
ciclosis, cyclosis
ciclotrón, cyclotron
cicuta, hemlock
ciego (del intestino), cecum
cielo, sky
cielo cubierto, overcast
ciempiés, centipede
ciénaga, swamp
ciencia, science
ciencias de la tierra, earth science
ciencias de la vida, life science
ciencias físicas, physical science
ciencias matemáticas, mathematical science
ciencias naturales, natural sciences
ciencias sociales, social science
cieno[1], silt
cieno[2], ooze
científico[1], scientific
científico[2], scientist
científico de materiales, materials scientist
cierto, certain
cigoto, zygote
cilindro, cylinder
cilindro vascular, vascular cylinder
cilio, cilium (pl. cilia)
cinabrio, cinnabar
cinc (Zn), zinc (Zn)
cincato, zincate
cinemática, kinematics
cinética, kinetics
cinta, ribbon
cinta métrica, tape measure
cintura, waist
cinturón de Kuiper, Kuiper belt
cinturón de Van Allen, Van Allen belt
circo, cirque
circón, zircon
circonio, zirconium
circuito, circuit
circuito abierto, open circuit
circuito cerrado, closed circuit
circuito en serie, series circuit
circuito en serie paralelo, series-parallel circuit
circuito exterior, external circuit

circuito integrado, integrated circuit
circuito interno, internal circuit
circuito paralelo, parallel circuit
circulación, circulation
circulación atmosférica, atmospheric circulation
circulación convectiva, convective circulation
circulación coronaria, coronary circulation
circulación portal hepática, hepatic portal circulation
circulación pulmonar, pulmonary circulation
circulación renal, renal circulation
circulación sanguínea, blood circulation
circulación simple, single circulation
circulación sistémica, systemic circulation
circular, circular
circulatorio, circulatory
círculo, circle
círculo mayor, great circle
círculo polar, polar circle
círculo polar antártico, Antarctic Circle
círculo polar ártico, Arctic Circle
circunferencia, circumference
circunscribir, circumscribe
cirro, cirrus
cirrocúmulo, cirrocumulus
cirrosis, cirrhosis
cirrostrato, cirrostratus
cirrus, cirrus, cirro
cirugía, surgery
cisteína, cysteine
citocinesis, cytokinesis
citocromo, cytochrome
citoesqueleto, cytoskeleton
citólisis, cytolysis
citología, cytology
citoplasma, cytoplasm
citosina, cytosine
citrato, citrate
cítrico, citrus
citronella, citronella
claridad, clarity
clarín, bugle
clarinete, clarinet
clase, class (biology)
clasificación[1], classification
clasificación[2], sorting, collating
clasificación horizontal, horizontal sorting
clasificar, classify
clave dicotómica, dichotomous key
clavícula, clavicle
clavo, nail
clima[1], climate
clima continental, continental climate
clima global, global climate
clima húmedo, humid climate
clima marino, marine climate
clima[2], weather
climatología, climatology
clítoris, clitoris
cloaca, cloaca
clon, clone
clonación, cloning

clonar, clone
clorar, chlorinate
clorato, chlorate
clorhidrato, hydrochloride
clorito, chlorite
cloro, chlorine
clorofila, chlorophyll
clorofluorocarbonos, chlorofluorocarbons
cloroformo, chloroform
clorohidrina, chlorohydrin
cloroplasto, chloroplast
cloruro, chloride
  cloruro de bario, barium chloride
  cloruro de calcio, calcium chloride
  cloruro de hidrógeno, hydrogen chloride
  cloruro de sodio, sodium chloride
  cloruro de vinilo, vinyl chloride
cnidarios, cnidarian
CNPT (condiciones normales de presión y temperatura), STP (standard condition for temperature and pressure)
coacervado, coacervate
coagular, coagulate
coágulo, clot, blood clot
cobalamina, cobalamin
cobalto, cobalt
cobertura de nubes, cloud cover
cobra, cobra
cobre, copper
cóccix, coccyx
cochinilla, scale insect
cociente, quotient
cóclea, cochlea
coco (bacteria), coccus
cocodrilo[1]**,** crocodile
cocodrilo[2]**,** crocodilian
código, code
  código ASCII, ASCII
  código binario, binary code
  código de colores, color code
  código de tripletes, triplet code
  código genético, genetic code
codo, elbow
codón, codon
coeficiente, coefficient
  coeficiente de expansión lineal, coefficient of linear expansion
  coeficiente de expansión volumétrica, coefficient of volume expansion
  coeficiente de fricción, coefficient of friction
  coeficiente de reparto, partition coefficient
cohesión, cohesion
cohesivo, cohesive
cohete, rocket
cohombro de mar, sea cucumber
coincidencia, coincidence
coito, (sexual) intercourse
cola, tail
colágeno, collagen
colapso por exceso de calor, heat exhaustion
colección, collection
cólera, cholera
colesterol, cholesterol
colibrí, hummingbird

coliflor, cauliflower
colina[1], drumlin
colina[2], hill
colineal, collinear
colisión, collision
  colisión efectiva, effective collision
  colisión inelástica, inelastic collision
colmena, hive
colmillo[1], fang
colmillo[2], tusk
coloide, colloid
colon, colon
  colon ascendente, ascending colon
  colon transverso, transverse colon
colonia, colony
  colonia de grajos, rookery
colonización, colonization
color, color
  color complementario, complementary color
  color gris azulado, iron blue
  color primario, primary color
  color secundario, secondary color
coloración, coloration
  coloración de advertencia, warning coloration
  coloración críptica, cryptic coloration
colorante, colorant
colorimetría, colorimetry
colorímetro fotoeléctrico, photoelectric colorimeter
columbio, columbium
columna, column
  columna vertebral, spinal column, vertebral column, spine
coma, coma
comadreja, weasel
combatir, combat
combinación, combination, arrangement
  combinación de colores, color scheme
  combinación directa, direct combination
combinar, combine, mix
combustible, fuel
  combustible fósil, fossil fuel
  combustible nuclear, nuclear fuel
combustión, combustion, burning
comer, eat
comercial, commercial
cometa, comet
cómodo, comfortable
comparar, compare
compartimento, compartment
compartir, share
compensación, trade-off
competencia, competition
  competencia interespecífica, interspecific competition
complejidad, complexity
complejo, complex
  complejo enzima-sustrato, enzyme-substrate complex
  complejos multicelulares, complex multicellular
complementario, complementary

complemento[1], complement
complemento[2], supplementation
componente, component
componente no perpendicular de un vector, nonperpendicular components of vector
componente no perpendicular, nonperpendicular component
componer, compose
comparación, comparison
comportamiento[1], behavior
comportamiento voluntario, voluntary behavior
comportamiento[2], behavioral
comportarse, behave
composición, composition
composición química, chemical composition
compresión, compression
comprimir, compress
compuesto[1], compound
compuesto cíclico, cyclic compound
compuesto estable, stable compound
compuesto inestable, unstable compound
compuesto inorgánico, inorganic compound
compuesto insaturado, unsaturated compound
compuesto intermetálico, intermetallic compound
compuesto orgánico, organic compound
compuesto ternario, ternary compound
compuesto[2], composite
compuesto[3], compounding
computadora, computer
computadora portátil, laptop
común, common
común divisor, common divisor
comunicación, communication
comunidad, community
comunidad clímax, climax community
comunidad de bienes, community property
cóncavo, concave
concentración, concentration
concepto, concept
concha, shell
conciencia, consciousness
concluir, conclude
conclusión, conclusion
conclusión de laboratorio, laboratory conclusion
concurrido, crowded
concusión, concussion
condensación, condensation
condensador, capacitor
condensar, condense
condición, condition
condición estándar, standard condition
condiciones normales de presión y temperatura (CNPT), standard condition for temperature and pressure (STP)
conducción, conduction
conducción electrolítica, electrolytic conduction

conducción iónica, ionic conduction
conducción metálica, metallic conduction
conducir, conduct
conducta, behavior
conductancia, conductance
conductividad, conductivity
conductividad calórica, thermal conductivity
conductividad eléctrica, electrical conductivity
conducto[1], duct
conducto biliar, bile duct
conducto colector, collecting duct
conducto deferente, vas deferens
conducto pancreático, pancreatic duct
conducto torácico, thoracic duct
conducto[2], channel
conducto auditivo, auditory canal
conductor, conductor
conejo, rabbit
conexión a tierra, grounding
conexión en serie, series connection
conexión en paralelo, parallel connection
confiable, reliable
confiar en, rely on
configuración, configuration
confinar, confine
conformación, conformation
conforme a, accordance
confortable, comfortable
congelación, freezing
congelado, frozen
congelar, freeze
congénito, congenital
conglomerado, conglomerate
congruente, congruent
conífera, conifer
conífero, coniferous
conjugación, conjugation
conjuntiva, conjunctiva
(en) conjunto, overall
conjunto[1], array
conjunto[2], set
conjunto vacío, empty set
conmoción, shock
conmutador, commutator
cono, cone
cono de ceniza, cinder cone
conocimiento, knowledge
consecuencias, consequences
conseguir[1], enlist
conseguir[2], attain, achieve
conservación, conservation
conservación de energía, conservation of energy
conservación de la masa, conservation of mass
conservación de la vida silvestre, wildlife conservation
conservación de los recursos de la Tierra, conservation of Earth resources
conservación del cambio, conservation of change
conservación del medio ambiente, conservation of environment

conservación del suelo, soil conservation
conservación forestal, forest conservation
conservante, preservative
conservar, conserve, retain
consideración del medio ambiente, environmental consideration
considerado, considered
considerar, regard
consiste en, consist of
consistente, consistent
consonancia, consonance
constante, constant
Constante de Boltzmann, Boltzmann constant
constante de equilibrio, equilibrium constant
constante de estabilidad, stability constant
constante de hidrólisis, hydrolysis constant
constante de ionización, ionization constant
constante del producto de solubilidad, solubility product constant
constante del producto iónico del agua, ion-product constant of water
constante gravitacional, gravitational constant
constante molal crioscópica, molal freezing point constant
constante molal ebulloscópica, molal boiling point constant
constelación, constellation
constipación, constipation
constituyente, constituent
construcción, construction
constructivo, constructive
constructor de barcos, ship builder
construir, construct
consulta, query
consumidor, consumer
consumidor de primer nivel, first-level consumer
consumidor de tercer nivel, third-level consumer
consumidor primario, primary consumer
consumidor secundario, secondary consumer
consumo de oxígeno, oxygen consumption
contabilidad de costos, cost accounting
contacto, contact
contacto tectónico, tectonic boundary
contagio, contagion
contagioso, contagious
contaminación, pollution
contaminación acústica, noise pollution
contaminación hídrica, water pollution
contaminación térmica, thermal pollution
contaminante, pollutant, contaminant
contaminar, contaminate
contenedor, container

contenedor de carbono, carbon-containing
contener, contain
contenido, content
continental, continental
continente, continent
Continente Antártico, Antarctic Continent
contracción, contraction
contracción muscular, muscle contraction
contraer, shrink
contraste, contrast
contratar, contract
contravenir, violate
contribuir, contribute
control, control
control biológico, biological control
control de calidad, quality control
control de la misión, Mission Control
control de natalidad, birth control
controversia entre natura y nurtura, nature and nurture controversy
convalecencia, convalescence
convección, convection
conveniente, convenient
convergencia, convergence
convergencia de la lente, converging lens
convergente, convergent
converger, converge
conversar, discuss
conversión, conversion
convertidor, converter
convertir, convert
convexo, convex
convulsión, convulsion
coordenada, coordinate
coordinación, coordination
copépodo, copepod
copolímero, copolymer
coral, coral
corazón, heart
corcho, cork
cordados, chordate
cordillera, cordillera
cordón costero, beach face
cordón litoral, barrier beach
cordón nervioso, nerve cord
cordón nervioso ventral, ventral nerve cord
cordón umbilical, umbilical cord
corindón, corundum
corión, chorion
cormo, corm
córnea, cornea
corneja, crow
corneta, bugle
cornezuelo, ergot
corola, corolla
corolario, corollary
corona[1], corona
corona[2], crown
coronario, coronary
corporal, somatic
corpóreo, somatic
corpúsculo, corpuscle
corredor, runner
correlación, correlation
correo electrónico, email
correspondiente, corresponding
corriente[1], current
corriente alterna (C.A.), alternating current (A.C.)
corriente continua, direct current
corriente de resaca, rip current

corriente eléctrica, electric current
corrientes superficiales, surface currents
corrientes de turbidez, turbidity currents
corriente[2], stream
corriente de descarga, stream discharge
(la) corriente del Golfo, Gulf Stream
corriente en chorro, jet stream
corrimiento al azul, blueshift
corrimiento de tierra, landslide
corroer, abrade
corrosión, corrosion
cortacésped, mower
cortacircuitos, circuit breaker
cortadora de césped, lawn mower
cortar, slit
corte, cutting
corteza[1], bark
corteza[2], cortex
corteza cerebral, cerebral cortex
corteza suprarrenal, adrenal cortex
corteza[3], crust
corteza continental, continental crust
corteza oceánica, oceanic crust
cortisona, cortisone
cortocircuito, short circuit
cosecante, cosecant
cosecha, harvesting
coseno, cosine
coser, sewing
cósmico, cosmic
cosmología, cosmology
cosmopolitas, cosmopolites
cosmos, cosmos
costa, coast
costa oceánica, coastal ocean
costilla, rib
costo-beneficio de las compensaciones, cost-benefit tradeoffs
cotangente, cotangent
cotejar, collate
cotiledón, cotyledon
cotización, quotation
covalencia, covalence
covalente, covalent
cráneo, skull
cráter, crater
cráter de impacto, impact crater
creación, foundation
crear, create
crecer, grow
crecimiento, growth
crecimiento exponencial, exponential growth
crecimiento logístico, logistic growth
cresol, cresol
cresta[1], crest
cresta[2], ridge
creta, chalk
Cretáceo, Cretaceous
cretinismo, cretinism
cría, breeding
crinoideo, crinoid
criogenia, cryogenics
crisálida[1], pupa
crisálida[2], chrysalis
crisis energética, energy crisis
crisol, crucible

crisol de sinterizado, sintered crucible
cristal, crystal
cristal de roca, flint glass
cristal líquido, liquid crystal
cristal semilla, seed crystal
cristalería, glassware
cristalino, crystalline
cristalización, crystallization
cristalización al vacío, vacuum crystallization
cristalizador de vacío, vacuum crystallizer
criterios dentro de las limitaciones, criteria within constraints
crítico, critical
cromático, chromatic
cromátida, chromatid
cromatina, chromatin
cromato de plomo, lead chromate
cromatografía, chromatography
cromo, chromium
cromo verde, chrome green
cromósfera, chromosphere
cromosoma, chromosome
cromosoma sexual, sex chromosome
cromosoma X, x-chromosome
cromosoma Y, y-chromosome
cronológico, chronological
cronómetro, chronometer, stopwatch
croquis, sketch
crotalino, pit viper
crujiente, crisp
crustáceo, crustacean
Cruz del Sur, Southern Cross
cruzamiento, crossbreeding, interbreeding
cruzar, crossbreed
ctenóforo, comb jelly
cuadernillo, booklet
cuadrado, square
cuadrangular, quadrangular
cuadrante, quadrant
cuadrático, quadratic
cuádriceps, quadriceps
cuadrilátero, quadrilateral
cuadrúpedo, quadruped
cualitativo, qualitative
cuantitativo, quantitative
cuantización, quantize
cuanto, quantum
cuantos de energía, quanta
cuarcita, quartzite
cuarentena, quarantine
cuarto, quart
cuarto de luna, quarter moon
cuarzo, quartz
cubeta de ondas, ripple tank
cúbico, cubic
cubierta[1]**,** cover
cubierta[2]**,** deck
cúbito, ulna
cubito de hielo, ice cube
cubo, cube
cubreobjetos, coverslip
cucaracha, cockroach, roach
cuchara, spoon
cuello, neck
cuello volcánico, volcanic neck
cuello uterino, cervix
cuenca, basin

cuenca de drenaje, drainage basin, water shed
cuenca de deposita- ción, depositional basin
cuerda[1], chord
cuerda[2], cord, rope, string
cuerda de arco, bowstring
cuerdas vocales, vocal cords
cuerno, antler, horn
cuerpo celular, cell body
cuerpo iluminado, illuminated body
cuerpo lúteo, corpus luteum
cueva, cave
cuidado parental, parental care
cuidadosamente, painstakingly
culebra, snake
culombio, coulomb
cultivar, culture
cultivo[1], culture
cultivo abierto, open field system
cultivo de tejidos, tissue culture
cultivo en semillero, seedling
cultivo[2], crop
cultivo en contorno, contour ploughing, contour plowing
cultivo en fajas, strip cropping
cultivos de cobertura, cover crop
cumbre, mountain peak
cúmulo, cumulus
cúmulo abierto, open cluster
cumulonimbo, cumulonimbus
cúmulus, cumulus
cuña, wedge
cuñete, keg
cúpula geodésica, geodesic dome
cura, cure
curio, curie, curium
curioso, curious
curso, trend, course
curso de acción, course of action
curva, curve
curva de solubilidad, solubility curve
curva de nivel, contour line, isoline
curva normal, bell curve
curvar, bend
curvatura[1], bent
curvatura[2], curvature
cutáneo, cutaneous
cutícula, cuticle
cyton, cyton

D

daltónico, color blind
daltonismo, color blindness
dañar, harm
dañino, harmful
daño, damage
dar, give
dar líneas aerodiná- micas, streamline
dardo, dart
Darwinismo, Darwinism
datación por carbono, carbon dating
datación relativa, relative dating

datos, data
datos atmosféricos, atmospheric data
datos científicos, scientific data
datos y cifras, Facts and Figures
DDT, DDT
deambular, meander
debajo[1], beneath
debajo[2], under
debate, debate
deber[1], owe
deber[2], duty
deber[3], due
deca-, deka-, deca-
decadencia, decay
decágono, decagon, decapod
decámetro, dekameter
deci-, deci-
decibel, decibel
decibelio, decibel
decimal, decimal
decímetro, decimeter
(de) décimo, deci
decisiones informadas, informed decisions
declaración, statement
declaración de impacto ambiental, environmental impact statement
declinación, declination
declinación magnética, magnetic declination
decrepitación, decrepitation
dedo, finger
dedo anular, ring finger
dedo del corazón, middle finger
dedo del pie, toe
dedo índice, index finger
dedo meñique, little finger
deducción, deduction
defecación, defecation
defecto, fault
defecto de masa, mass defect
defectuoso, defective
defender, defend
deficiente, deficient
déficit, deficit
definición, definition
definición conceptual, conceptual definition
definición operativa, operational definition
definir, define
definitivo, definite
deflector, baffle
defoliación, defoliation
deforestación, deforestation
deformación, deformation
deformación cortical, crustal down-warping
deformación de las rocas, deformation of rocks
degeneración, degeneration
delfín, dolphin
delgado, thin
delicuescencia, deliquescence
delta, delta
demandar, demand
demografía, demography
demostración, demonstration
demostrar, demonstrate
dendrita, dendrite
dendrocronología, dendrochronology
denominador, denominator

denominador común, common denominator
densidad, density
densidad óptica, optical density
dentición, dentition
dentina, dentin
dentro de, within
depender, depend
depilar, waxing
deposición, deposition
depositar, settle
depósito, deposit
depósito mineral, ore deposit
depredación, predation
depredador[1]**,** predator
depredador[2]**,** predatory
depresión, depression
depresión del punto de congelación, freezing point depression
depresión de bulbo húmedo, wet-bulb depression
depresivo, depressant
derecho, standing
deriva, drift
deriva continental, continental drift
deriva litoral, longshore drift
derivada, derivative
derivado, derived
derivar, derive
dermatoesqueleto, exoskeleton
dermis, dermis
derramar[1]**,** spill
derramar[2]**,** pour
derrame, spill
derrame cerebral, stroke
desaceleración, deceleration
desafío, challenge
desalación, desalinization
desalinización, desalinization
desaminación, deamination
desarrollar, develop
desarrollo, development
desarrollo interno, internal development
desastre, disaster
desastre perjudicial, perjudicial disaster
desastre natural, natural disaster
descarga, discharge
descargar, dump
descendencia, offspring
descender, descend
descenso, descent
descomposición, decomposition, decay
desconocido, unknown
describir, describe
descubierto, bare
descubrimiento[1]**,** discovery
descubrimiento[2]**,** strike
descubrir, discover
desdentado, edentate
deseable, desirable
desechos, debris
desechos metabólicos, metabolic waste
desembocadura, outlet
desequilibrar, unbalance
desertización, desertification
desfiladero, gorge, gully
desgarrar, rip
desgarrarse, tear
desgaste, weathering
desgaste físico, physical weathering
desgaste mecánico, mechanical weathering

desgaste químico, chemical weathering
deshidratación, dehydration
deshidrogenación, dehydrogenation
deshidrogenasa, dehydrogenase
deshielo, deglaciation
desierto, desert
desierto urbano, urban desert
desigual[1], unequal
desigual[2], rough
desintegración radiactiva, radioactive decay
desintegrar, disintegrate
desintoxicación, detoxication
deslizar, slide
deslumbramiento por la nieve, snow blindness
desmayarse, faint
desnaturalización, denaturation
desnitrificación, denitrification
desnivel, ramp
desnudo, bare, naked
desnutrición, malnutrition
desoxirribosa, deoxyribose
desplazamiento, displacement
desplazamiento de las series, displacement series
desplazamiento de sedimentos, displacement sediments
desplazamiento hacia el rojo, redshift
desplazar, displace
desplazar el cursor, scroll (computer)
desproporcionada, disproportionate
destilación, distillation
destilación al vacío, vacuum distillation
destilación a vapor, steam distillation
destilación fraccionada, fractional distillation
destilado, distillate
destornillador, screwdriver
destrucción, destruction
destructivo, destructive
destruir, destroy
desventaja, disadvantage
desviación, deviation
desviar, deflect
desviarse, deviate
detección, detection
detectar, detect
detector, detector
detener, halt
detergente, detergent
determinar, determine
detiene, stops
detrás, behind
detrito, detritus
deuda de oxígeno, oxygen debt
deuterio, deuterium
deuterón, deuteron
dextrógiro, dextrorotatory
dextrosa, dextrose
día, day
día sideral, sideral day
día solar aparente, apparent solar day
día solar medio, mean solar day
diabetes, diabetes
diafragma, diaphragm
diagnóstico, diagnosis

diagonal, diagonal
diagrama, diagram
diagrama circular, pie chart
diagrama de flujo, flow chart, flow diagram
diagrama de pedigrí, pedigree chart
diagrama de Punnett, Punnett square
diagrama de puntos, dot diagram
diagrama indicador, diagram key
diálisis, dialysis
diamante, diamond
diamantes industriales, industrial diamonds
diámetro, diameter
diámetro aparente planetario, apparent planetary diameter
diámetro aparente, apparent diameter
diapasón, tuning fork
diarrea, diarrhea
diástole, diastole
diastrofismo, diastrophism
diatomea, diatom
diborano, diborane
dibujar, draw
dicloruro de etileno, ethylene dichloride
dicotiledóneas, dicotyledon
dicotómica, dichotomous
dieléctrico, dielectric
diente (pl. dientes), tooth (pl. teeth)
diente de leche, milk tooth
dientes (sing. diente), teeth (sing. tooth)
dieta, diet
dieta balanceada, balanced diet
dietilamina, diethylamine
diferencia, difference
diferencia de potencial eléctrico, difference in electric potential
diferenciación, differentiation
diferentes, various
diferir, differ
difracción, diffraction
difracción de una sola rendija, single-slit diffraction
difracción de doble rendija, double slit diffraction
difractar, diffract
difteria, diphtheria
difuminar, blur
difusión, diffusion
difusión facilitada, facilitated diffusion
difusión térmica, thermal diffusion
difuso, diffuse
digerir, digest
digestión, digestion
digestión extracelular, extracellular digestion
digestión intracelular, intracellular digestion
digital, digital
dígito, digit
dígito binario, binary digit
dígitos significativos, significant digits
dihíbrido, dihybrid
dihidroxi, dihydroxy
dilatación, dilation

dilatación de los vasos sanguíneos, dilation of blood vessel
dilatar, dilate
dilución, dilution
diluente, diluent
diluir, dilute
dimensión, dimension
dímero, dimer
dimethylketone, dimethylketone
diminuto, tiny
dimorfismo, dimorphism
dina, dyne
dinámica, dynamics
dinámico, dynamic
dinamita, dynamite
dinamo (dínamo), dynamo
dinamómetro, spring scale
dinitrobenceno, dinitrobenzene
dinosaurio, dinosaur
diodo, diode
diodo emisor de luz (LED), light-emitting diode (LED)
dioico, dioecious
diolefin, diolefin
diorita, diorite
dióxido, dioxide
dióxido de azufre, sulfur dioxide
dióxido de carbono (CO_2), carbon dioxide (CO_2)
dióxido de carbono atmosférico, atmospheric carbon dioxide
dióxido de cloro, chlorine dioxide
dióxido de manganeso, manganese dioxide
dipéptido, dipeptide
dipolo, dipole
dique, dam, levee, dike
diques naturales, natural levees
dirección[1], direction
dirección[2], trend
directamente, directly
directo, direct
directorio, directory
directriz, directrix
dirigir, manage
disacárido, disaccharide
disco[1], disk
disco compacto, compact disk
disco duro, hard disk
disco flexible, floppy disk
disco[2], puck
disconformidad, disconformity, nonconformity
discontinuidad de Mohorovicic, Mohorovicic discontinuity
discordancia angular, angular unconformity
discordancia paralela, parallel unconformity
discriminación sexual, sexual discrimination
discutir, discuss
disecar, dissect
disección, dissection
diseccionar, dissect
diseñar, design
diseño, design
diseño experimental, experimental design
disentería, dysentery
disfasia, dysphasia
disipador de energía,

energy sink
disipar, dissipate
dislexia, dyslexia
dislocación, dislocation
disminución, decrease
disociación, dissociation
disolución, dissolution
disolvente, solvent
disolver, dissolve
disonancia, dissonance
disparador, trigger
dispersar, disperse
dispersión[1], dispersal
dispersión de semillas, seed dispersal
dispersión[2], dispersion
dispersión coloidal, colloidal dispersion
dispersión uniforme, uniform dispersion
dispersión[3], scattering
disponer, dispose
disponibilidad, availability
disponible, available
dispositivo, device
dispositivo termonuclear, thermonuclear device
disprosio, dysprosium
disquete, floppy disk
distancia, distance
distancia focal, focal length
distante, distant
distintivo, distinctive
distorsión, distortion
distorsionar, distort
distribución, distribution
distribución global, global distribution
distribución independiente, independent assortment
distribuir, distribute
distributario, distributary
distrofia muscular, muscle dystrophy
disulfuro de carbono, carbon disulfide
disyunción, disjunction, nondisjunction
diurético, diuretic
diurno, diurnal
divergencia, divergence
divergente, divergent
divergir, diverge
diversidad, diversity
diversidad de ecosistemas, ecosystem diversity
diversidad genética, genetic diversity
dividendo, dividend
dividir[1], divide
dividir[2], split
división, division
división celular, cell division
división celular mitótica, mitotic cell division
división citoplasmática, cytoplasmic division
división continental, continental divide
divisor[1], divisor
divisor[2], divider
divisoria continental, continental divide
doblar, bend
doble, double
doble enlace, double bond
doble fertilización, double fertilization

doble hélice, double helix
dodecaedro, dodecahedron
dodecágono, dodecagon
dolomita, dolomite
dominante, dominant
dominio[1], domain
dominio[2], dominance
donante, donor
donante de protones, proton donor
Doppler, Doppler
dormir, sleep
dorsal, dorsal
dorsal oceánica, mid-ocean ridge
dos veces, twice
dosel, canopy
drenaje, drainage
droga[1], drug
droga sulfa, sulfa drug
droga[2], dope
drogodependencia, drug dependence
drosophila, drosophila
drupa, drupe
dúctil, ductile
ductilidad, ductility
dueño de casa, homeowner
duna, dune
duodecimal, duodecimal
duódeno, duodenum
duración de la insolación, duration of insolation
duración de vida, life span
duración del día, day length
dureza, hardness
duro(a), hard

ebullición, boiling
eccema, eczema
eclipse, eclipse
eclipse de sol, solar eclipse
eclipse lunar, lunar eclipse
eclíptica, ecliptic
eco, echo
ecografía, ultrasonography
ecología, ecology
ecológica, ecological
ecológicamente, ecologically
economía, economics
económicamente, economically
económico[1], economical
económico[2], inexpensive
ecosfera, ecosphere
ecosistema, ecosystem
ectodermo, ectoderm
ecuación, equation
ecuación básica, basic equation
ecuación de Nernst, Nernst equation
ecuación de van't Hoff, van't Hoff equation
ecuación lineal, linear equation
ecuación química, chemical equation
ecuador, equator
Ecuador celeste, celestial equator
Edad de Piedra, Stone Age
edad geológica, geologic time
edema, edema
educación sexual, sex

education
efectivamente, effectively
efectivo, effective
efecto, effect
Efecto Compton, Compton effect
efecto Coriolis, Coriolis effect
efecto del ion común, common ion effect
efecto del calor, heat effect
efecto del Niño, El Niño effect
efecto Doppler, Doppler effect, Doppler shift
efecto secundario, side effect
efecto fotoeléctrico, photoelectric effect
efecto invernadero, greenhouse effect
efecto orográfico, orographic effect
efecto piezoeléctrico, piezoelectric effect
efector, effector
eferente, efferent
efervescencia, effervescence
eficacia, effectiveness
eficiencia, efficiency
eficiente, efficient
eflorescencia, efflorescence
egestión, egestion
Einstein, Einstein
einstenio, einsteinium
eje[1], axis
eje principal, principal axis
eje transversal, cross axis
eje x, x-axis
eje y, y-axis
eje z, z-axis
eje[2], axle
ejecutante, performer
ejemplo, example
ejercer, exert
ejercicio, exercise
elástica, elastic
elasticidad[1], elasticity
elasticidad[2], restoring force
elección, choice
electricidad, electricity
electricidad estática, static electricity
eléctrico, electric
electrocardiograma (ECG), electrocardiogram (ECG or EKG)
electrodeposición, electrodeposition
electrodinámica, electrodynamics
electrodo, electrode
electrodo de vidrio, glass electrode
electrodo normal de calomelanos, standard calomel electrode
electroencefalograma, electroencephalogram
electroforesis, electrophoresis
electroforesis en gel, gel electrophoresis
electroimán, electromagnet
electrólisis, electrolysis
electrolito, electrolyte
electrolito débil, weak electrolyte

electrolito fuerte, strong electrolyte
electrolizar, electrolyze
electromagnético, electromagnetic
electromagnetismo, electromagnetism
electromotriz, electromotive (EMF)
electrón, electron
electronegativo, electronegative
electrónico, electronic
electronvoltio, electronic volt
electropositivo, electropositive
electroquímica, electrochemistry
electroquímico, electrochemical
electroscopio, electroscope
electrovalencia, electrovalence
elefante, elephant
elefantiasis, elephantiasis
elegir, choose
elemento, element
elemento de transición, transition element
elemento traza, trace element
elemento radiactivo, radioactive element
elemento sólido inorgánico, inorganic solid element
elementos ligeros, light elements
elementos transuránicos, transuranic element
elevación, elevation
elevación del punto de ebullición, boiling point elevation
elevador, elevator
eliminación, disposal
eliminar, eliminate
elipse, ellipse
elíptico, elliptical
elongación, elongation
embalaje, packing
embalse[1], dam
embalse[2], reservoir
embarazo, pregnancy
embarcadero, jetty
embolia, embolism
émbolo, embolus
embriología, embryology
embrión, embryo
embrionario, embryonic
embudo, funnel
emergencia[1], emergency
emergencia[2], emergence
emerger, surface
emersión continental, continental rise
emigración, emigration
emisión, emission
emisión alfa, alpha emission
emisión estimulada, stimulated emission
emisora, transmitter
emitir[1], emit
emitir[2], cast
empacadora, gasket
empalizada, palisade
empalmar, splice
empapar, soak
empaque, gasket
empaquetadura, gasket
empírico, empirical
empuje, thrust
emú, emu

emulsificación, emulsification
emulsión, emulsion
enana blanca, white dwarf
enana marrón, brown dwarf
enana negra, black dwarf
enana parda, brown dwarf
enana roja, red dwarf
enanismo, dwarfism
enanismo hipofisario, pituitary dwarfism
enano, dwarf
encadenar, catenate
encargar, set
encefalitis, encephalitis
encefalitis letárgica, sleeping sickness
encender, ignite
encendido, ignition
encerar, waxing
enchufe, socket
encía[1], gingiva
encía[2], gum
encoger, shrink
encontrarse con, encounter
encuesta, survey
endémico, endemic
enderezar, straighten
endocrinología, endocrinology
endodermo, endoderm
endoesqueleto, endoskeleton
endogamia, inbreed, inbreeding
endometrio, endometrium
endoparásitos, endoparasite
endorfina, endorphin
endospermo, endosperm
endospora, endospore
endotérmico, endothermic
enea, cattail
energética, energetic
energía, energy
energía atómica, atomic energy
energía cinética, kinetic energy
energía de activación, activation energy
energía de ionización, ionization energy
energía eléctrica, electrical energy
energía en onda larga, long-wave energy
energía geotérmica, geothermal energy
energía hidroeléctrica, hydroelectric power
energía interna, internal energy
energía libre, free energy
energía mareomotriz, tidal energy
energía mecánica, mechanical energy
energía nuclear, nuclear energy
energía potencial, potential energy
energía potencial nuclear, nuclear potential energy
energía química, chemical energy
energía radiante, radiant energy
energía solar, solar energy, solar power
energía térmica, heat energy, thermal energy
enfermedad[1], ailment
enfermedad[2], disease

enfermedad cardiovascular, cardiovascular disease
enfermedad de Alzheimer, Alzheimer's disease
enfermedad de deficiencia, deficiency disease
enfermedad de las vacas locas, mad cow disease
enfermedad de Lyme, Lyme disease
enfermedad de Parkinson, Parkinson's disease
enfermedad de Tay-Sachs, Tay-Sachs disease
enfermedad de transmisión sexual (ETS), sexually transmitted disease (STD)
enfermedad genética, genetic disease
enfermedad mental, mental illness
enfermedad venérea, venereal disease
enfermedad viral, viral disease
enfisema, emphysema
enganchar, hook
engranaje, gear
engranar, mesh
enlace[1]**,** bond
enlace covalente, covalent bond
enlace de alta energía, high energy bond
enlace de hidrógeno, hydrogen bond
enlace iónico, ionic bonding, ionic bond
enlace metálico, metallic bond
enlace peptídico, peptide bond
enlace polar, polar bond
enlace químico, chemical bond
enlace triple, triple bond
enlace[2]**,** link
enmarcar, frame
enorme, enormous, huge
enriquecer, enrich
enriquecimiento, enrichment
enrollar la lengua, tongue rolling
enrutador, router
ensanchar, widen
ensayo, trial
ensayo de la mancha de grasa, spot test for fat
entalpía, enthalpy
entalpía de fusión, heat of fusion
entalpía de solución, heat of solution
entalpía estándar de formación, standard heat of formation
entender, see
entomología, entomology
entorno, environment
entorno físico, physical environment
entrada, input
entrar, enter
entrar en erupción,

erupt
entregar, deliver
entrelazar, interlock
entropía, entropy
enumerar, list
envejecimiento, aging
envenenar, poison
envergadura, wingspan
enzima, enzyme
eón, eon
eperlano, smelt
epicentro, epicenter
epicentro del terremoto, earthquake epicenter
epicótilo, epicotyl
epidemia, epidemic
epidemiología, epidemiology
epidermis, epidermis
epidídimo, epididymis
epifita, epiphyte
epiglotis, epiglottis
epilepsia, epilepsy
epinefrina, epinephrine
epitelial, epithelial
epitelio, epithelium
época[1], epoch
época[2], time
epoxi, epoxy
equidistante, equidistant
equidna, echidna
equilibrante, equilibrant
equilibrio[1], balance
equilibrio del medio ambiente, environmental balance
equilibrio de sedimentación, sedimentation balance
equilibrio[2], equilibrium
equilibrio físico, physical equilibrium
equilibrio interrumpido, punctuated equilibrium
equilibrio térmico, thermal equilibrium
equino[1], equine
equino[2], sea urchin
equinoccio, equinox
equinoccio de otoño, autumnal equinox
equinoccio de primavera, vernal equinox
equinodermo, echinoderm
equipo, equipment
equivalente, equivalent
era, era
Era Mesozoica, Mesozoic Era
Era Precámbrica, Precambrian Era
Era Cenozoica, Cenozoic Era
Era Paleozoica, Paleozoic Era
erbio, erbium
erección, erection
ergio, erg
ergonomía, ergonomics
erigir, erect
eritrocito, red blood cell
erizo de mar, sea urchin
erosión, erosion
erosión de la línea lateral y la cabeza, head erosion
erosión del suelo, soil erosion
erosión eólica, wind erosion
erosión física, physical weathering
erosión lateral, lateral erosion

erosión y depósito en sistemas, erosional–depositional system
erosionar, abrade
erráticas, erratics
error, error
error de medida, error extent
erudito, learned
erupción solar, solar flare
escafandra autónoma, scuba
escala, scale
escala absoluta, absolute scale
escala Celsius, Celsius scale
escala de ebullición, boiler scale
escala de Mercalli modificada, modified Mercalli scale
escala de Richter, Richter scale
escala de temperatura, temperature scale
escala global, global scale
escalar, scalar
escaleno, scalene
escandio, scandium
escápula, scapula
escarabajo[1], beetle
escarabajo[2], scarab
escarabajo[3], roach
escarcha, frost
escarlatina, scarlet fever
escarpa, escarpment
escarpada, pavement
escasez, scarcity
escaso, scarce
Escherichia coli (E. coli), Escherichia coli (E. coli)
esclereidas, sclereid
esclerosis múltiple, multiple sclerosis
esclerótica, sclera
escoliosis, scoliosis
escorbuto, scurvy
escoria, scoria
escorpión, scorpion
escroto, scrotum
escuchar, hear
escudo, shield
escudo de cono, shield cone
escudo de la capa de ozono, ozone shield
escurrir, seep
esencia[1], core
esencia[2], essence
esencialmente, essentially
esfagno, sphagnum
esfalerita, sphalerite
esfera, sphere
esfera celeste, celestial sphere
esférica, spherical
esferoide, spheroid
esferoide oblato, oblate spheroid
esfínter, sphincter
esfínter pilórico, pyloric sphincter
esfuerzo, effort
esguince, sprain
esker, esker
eslabón, link
esmalte, enamel
esmeralda, emerald
esmog, smog
esófago, esophagus, gullet
espacio, space
espacio aéreo, air space
espalda, back

especiación, speciation
especialización celular, cell specialization
especializar, specialize
especie, species
  especie dominante, dominant species
  especie exótica, alien (ecology)
  especie invasiva, invasive species
  especie pionera, pioneer species
  especies en vías de extinción, endangered species
especificidad, specificity
específico, specific
especimen, specimen
espectacular, spectacular
espectro, spectrum
  espectro continuo, continuous spectrum
  espectro de absorción, absorption spectrum
  espectro de emisión, emission spectrum
  espectro electromagnético, electromagnetic spectrum
  espectro de línea brillante, bright line spectrum
  espectro visible, visible spectrum
espectrógrafo, spectrograph
  espectrógrafo de masas, mass spectograph
espectrometría de absorción atómica, atomic absorption spectrometry
espectrómetro, spectrometer
espectros, spectra
espectroscopia, spectroscopy
  espectroscopia infrarroja, infrared spectroscopy
espectroscopio, spectroscope
espejismo, mirage
espejo, mirror
  espejo cóncavo, concave mirror
  espejo convexo, convex mirror
esperanza de vida, life expectancy
esperar[1], expect
esperar[2], wait
esperlán, smelt
esperma, sperm
  esperma de ballena, spermaceti
espermaceti, spermaceti
espermátide, spermatid
espermatocito, spermatocyte
espermatofita, spermatophyte
espermatogénesis, spermatogenesis
espermatozoario, spermatozoon
espermatozoide, spermatozoon
  espermatozoides del conducto deferente, sperm duct
espermátulo, spermatozoon
espetar, spit
espina[1], thorn
espina[2], spine
  espina dorsal, backbone, spinal cord

espina[3] (pez), bone
espinal, spinal
espinazo, spine
espinilla, pimple
espiráculo, spiracle
espiral, spiral
espliego, spike
espoleta, wishbone
espolón, dewclaw
espolvorear, sprinkle
esponja, sponge
espontáneamente, spontaneously
espora, spore
esporangio, sporangium
esporogénesis, sporogenesis
esporulación, sporulation
espuma, foam
esputo, sputum
esqueleto, skeleton, skeletal system
esquemática, schematic
esquisto, schist, shale
esquizofrenia, schizophrenia
estabilidad, stability
estabilización, stabilization
estabilizador, stabilizer
estable, stable
establecer, establish
establecimiento, establishment
estaca, stake
estación, season
  estación de modelo, station model
estacionalmente, seasonally
estacionario[1], stable
estacionario[2], stationary
estadística, statistics
estado, state
  estado de agregación de la materia, state of matter
  estado de equilibrio, steady state
  estado de excitación, excited state
  estado de vapor, vapor state
  estado energético, energetic state
  estado fundamental, ground state
  estado sólido, solid state
estafilococo, staphylococcus
estalactita, stalactite
estalagmitas, stalagmites
estambre, stamen
estaminífero, staminate
estampido sónico, sonic boom
estancado, stagnant
estándar, standard
estánico, stannic
estannoso, stannous
estaño, tin
estanque, pond
estante, shelf
estático, static
estepa, steppe
éster, ester
estereoisómero, stereoisomer
estereoquímica, stereochemistry
estereoscopio, stereoscope
esterificación, esterification
estéril, infertile, sterile
esterilidad, sterility
esterilización, sterilization
esterilizante, sterilizing
esternón, sternum
esteroide, steroid
esterol, sterol

estetoscopio, stethoscope
estibador, dockhand
estiércol líquido, slurry
estigma, stigma
estilo, style
estimación, estimation
estimulante, stimulant
estimular, stimulate
estímulo, stimulus (pl. stimuli)
estípula, stipule
estiramiento, stretch
estivación, estivation
estivar, estivate
estolón, stolon
estoma[1], stoma
estoma[2], stomata
estómago, stomach
estomas, stomate
estrabismo, squint
estrategia, strategy
estratificación, stratification, bedding
  estratificación cruzada, cross-bedding
estratificada, stratified
estrato[1], strato
estrato[2], stratum
estrato[3], stratus (pl. strata)
estratocúmulus, stratocumulus
estratos, strata (sing. stratus)
  estratos plegados, folded strata
  estratos inclinados, tilted strata
  estratos sedimentarios, sedimentary strata
estratósfera, stratosphere
estrechamente, tightly
estrecho, strait
estrella, star
  estrella binaria, binary star
  estrella de neutrones, neutron star
  estrella doble, binary star
  estrella enana, dwarf star
  estrella fija, fixed star
  estrella fugaz, fireball
  estrella fugaz, shooting star
  estrella gigante, giant star
  Estrella Polar, North Star
  estrella variable, variable star
  estrellas circumpolares, circumpolar stars
estrella de mar, starfish
estrellar, crash, smash
estrepitoso, loudness
estreptococo, streptococcus
estreptomicina, streptomycin
estrés, stress
estriación, striation
estribo, stapes
estricnina, strychnine
estridente, loudness
estro, estrus
estroboscópico, strobe
estrógeno, estrogen
estroma, stroma
estroncio, strontium
estropear, mar
estructura, structure
  estructura cristalina, crystal structure, crystalline structure

estructura de anillo, ring structure
estructura distorsionada, distorted structure
estuario, estuary, firth
estupefaciente, narcotic drug
esturión, sturgeon
etanal[1], acetaldehyde
etanal[2], ethanal
etano, ethane
etanol, ethanol
etanolamina, ethanolamine
etapas, stages
eteno, ethene
éter, ether
éter etílico, ethyl ether
éter de polivinilo, polyvinyl ether
ética, ethical
etilamina, ethylamine
etilbenceno, ethylbenzene
etilenglicol, ethylene glycol
etileno, ethylene
etilo, ethyl
etiología, etiology
etiqueta, label
etiquetar, label
eucalipto, eucalyptus
eucariota, eukaryote, eukaryotic
euglena, euglena
europio, europium
eutéctica, eutectic
eutrofización, eutrophication
evaluación justa, fair test
evaluar, evaluate
evaporación, evaporation
evaporar, evaporate
evaporitas, evaporites
evapotranspiración, evapotranspiration
evapotranspiración potencial, potential evapotranspiration
evapotranspiración real, actual evapotranspiration
evento, event
eventualmente, eventually
evidencia, evidence
evidencia fósil, fossil evidence
evidenciar, evidence
evolución, evolution
evolución convergente, convergent evolution
evolucionar, evolve
evolutivo, evolutionary
exactitud, accuracy
exageración, magnification
examen, test, examination
examinar, overlook
excavadora, steam shovel
excavar, hollow
excentricidad, eccentricity
excepción, exception
excepcional, rare
excesivo, excessive
exceso, excess
exceso de concentración en la superficie, surface concentration excess
excitación, excitation
excluir, exclude
exclusivo, exclusive
excreción, excretion
excretar, excrete
excretor, excretory
excursionista, hiker
exfoliación, exfoliation
exhalación, exhalation

exhibir, exhibit
existir, exist
existencia, existence
exogamia, outbreeding
exón, exon
exósfera, exosphere
exotérmico, exothermic
expandir, expand
expansión, expansion
expansión del fondo marino, sea floor spreading
expansión del fondo marino, ocean-floor spreading
expansión térmica, thermal expansion
expectorante, expectorant
experimentar, experiment, experimenting
experimento, experiment
experimento controlado, controlled experiment
expiración, expiration
explicación[1], explanation
explicación[2], account
explicar, explain
exploración sísmica, seismic exploration
explorar[1], explore
explorar[2], surf
explosión, explosion
explosivo, explosive
explosivo de gran potencia, high explosive
explosivo inicial, initiating explosive
explotación agrícola, husbandry
explotación forestal, logging
explotación ganadera, husbandry
explotación de la fauna, exploitation of fauna
exponencial, exponential
exponente, exponent
exponer, expose
exposición, exposure
expresado, expressed
expresar, express
expresión, expression
expresión del producto de solubilidad, solubility product expression
expreso, express
expuesto, exposed
expulsar, expel
extensor, extensor
exterior, outward
externo, external
extinción, extinction
extinción de la fauna, extinction of fauna
extinguido, extinct
extinguir, extinct
extinguirse, died off
extirpación, ablation
extracción, extraction
extraer, extract
extrapolación, extrapolation
extrapolar, extrapolate
extraterrestre, extraterrestrial
extremadamente, extremely
extremo[1], extreme
extremo[2], endpoint
extrusión, extrusion

fabricación en serie, mass production
fabricar, manufacture
faceta, facet
facetada, faceted
fácilmente, readily
facticio, factice
factor, factor
  factor abiótico, abiotic factor
  factor biótico, biotic factor
  factor común, common factor
  factor de conservación, conservation factor
  factor de enfriamiento del viento, wind-chill factor
  factor frío del viento, wind chill factor
  factor limitante dependiente de densidad, density-dependant limiting factor
  factor limitante independiente de densidad, density-independent limiting factor
  factor Rhesus (factor Rh), Rhesus factor (Rh factor)
  factor variable, variable factor
factorial, factorial
fagocito, phagocyte
fagocitosis, phagocytosis
faja, strip
falanges, phalanges
falla[1]**,** rift
falla[2]**,** fault
  falla inversa, reverse fault
  falla lateral, lateral fault
  falla normal, normal fault
fallar, fail
familia (biología), family (biology)
familia del nitrógeno, nitrogen family
fango, ooze
Faraday, Faraday
faradio, farad
faringe, pharynx
farmacología, pharmacology
faro, headlamp
fase, phase
  fase de reposo (interfase), resting stage (interphase)
  fase del gas, gas phase
  fase de equilibrio, phase equilibrium
  fase líquida, liquid phase
  fase sólida, solid phase
fatiga, fatigue
  fatiga muscular, muscle fatigue
fauna, fauna
  fauna clímax, climax fauna
fecundación, conception
feldespato, feldspar
felino, feline
félsica, felsic
fémur, femur
fenil, phenyl

fenilalanina, phenylalanine
fenilcetonuria (PKU), phenylketonuria (PKU)
fenilo, phenyl
fenol, phenol
fenolftaleína, phenolphthalein
fenólicos, phenolic
fenómeno, phenomenon (pl. phenomena)
fenómeno celeste, celestial phenomenon
fenómenos físicos, physical phenomena
fenotipo, phenotype
fermentación, fermentation
fermio, fermium
feromona, pheromone
férrico, ferric
ferroaleaciones, ferroalloy
ferroso, ferrous, ferric
fértil, fertile
fertilización, fertilization
fertilización cruzada, cross-fertilization
fertilización externa, external fertilization
fertilización in vitro, in vitro fertilization
fertilización interna, internal fertilization
fertilizar, fertilizer
fetal, fetal
feto, fetus
fibra, fiber
fibra de vidrio, fiberglass
fibra del huso, spindle fiber
fibra leñosa, woody fiber
fibra nerviosa, nerve fiber
fibra óptica, fiber optics, optical fiber
fibras nerviosas sensoriales, sensory nerve fiber
fibrilación, fibrillation
fibrina, fibrin
fibrinógeno, fibrinogen
fibrosis quística, cystic fibrosis
ficción, fiction
ficoeritrina, phycoerythrin
fiebre, fever
fiebre aftosa del ganado, foot-and-mouth disease
fiebre amarilla, yellow fever
fiebre del heno, hay fever
fiebre manchada de las Montañas Rocosas, Rocky Mountain spotted fever
fiebre reumática, rheumatic fever
fiebre tifoidea, typhoid fever
fijación del carbono, carbon fixation
fijar, set
fijo, set, fixed
filamento, filament
fileras, spinneret
filón, seam
filósofo, philosopher
filtración, filtration
filtrado, filtered
filtrar[1], seep
filtrar[2], filter
filtrar[3], filtrate
filtrar[4], leak
filtro, filter

filum, phylum
finito, finite
fino, fine
fiordo, fjord, fiord
firme, firm
física, physics
  física clásica, classical physics
  física nuclear, nuclear physics
físico[1], somatic
físico[2], physical
físico[3], physicist
  físico teórico, theoretical physicist
fisicoquímica, physical chemistry
fisiología, physiology
fisión, fission
  fisión binaria, binary fission
  fisión nuclear, nuclear fission
fisioterapia, physiotherapy
fisisorción, physiosorption
fisura, fissure
fitoplancton, phytoplankton
flagelados, flagellate
flagelo, flagellum (pl. flagella)
flamenco, flamingo
flecha, arrow
flema, phlegm
flexor, flexor
floema, phloem
flogisto, phlogiston
flor[1], flower
  flor estaminada, staminate flower
flor[2], blossom
flora, flora
  flora clímax, climax flora
floración de algas, algal bloom
florecer, blossom
florescencia, florescence
flotabilidad, buoyancy
flotación, flotation
flotante, buoyant
flotar, float
fluido, fluid
  fluido de tejidos, tissue fluid
flujo[1], flux
flujo[2], flow
  flujo ascendente, ascending flow
  flujo de energía, flow of energy, energy flow
  flujo de lava, lava flow
  flujo descendente, descending flow
  flujo laminar, laminar flow
  flujo luminoso, luminous flux
  flujo magnético, magnetic flux
  flujo turbulento, turbulent flow
  flujo venoso, venous flow
flúor, fluorine
fluoración, fluoridation
fluorescencia, fluorescence
fluorescente, fluorescent
fluorita, fluorite, fluorspar
fluorizar, fluoridate
fluorocarbono, fluorocarbon
fluoroscopia, fluoroscopy
fluoruro, fluoride
  fluoruro de hidrógeno, hydrogen fluoride
  fluoruro de sodio,

sodium fluoride
foca, seal
focal, focal
foco, focus (pl. foci)
foco principal, principal focus
focos, foci
foliada, foliated
folículo, follicle
folículo piloso, hair follicle
foliolo, leaflet
folleto, booklet
fondo cósmico, cosmic background
fórceps, forceps
forma, shape, form
forma de los cristales, crystal shape
forma de vida, life-form
forma del canal, channel shape
formación, formation
formación de nubes, cloud formation
formación rocosa, rock formation
formación de roca ígnea, igneous rock formation
formaldehído, formaldehyde
formar, shape, form
fórmula, formula
fórmula de la masa, formula mass
fórmula configuracional, configurational formula
fórmula empírica, empirical formula
fórmula estructural, structural formula
fórmula molecular, molecular formula
forraje, roughage
fosfato, phosphate
fosfato de calcio, calcium phosphate
fosfato tricresilo, tricresyl phosphate
fosforescencia, phosphorescence
fosforilación, phosphorylation
fósforo[1], phosphor
fósforo[2], phosphorus
fósil, fossil
fósil índice, index fossil
fosilización, fossilization
fosos de alquitrán, tar pit
fotocopiar, Xerox
fotodegradable, photodegradable
fotoeléctrico, photoelectric
fotografía, photography
fotografía estroboscópica, strobe photography
fotólisis, photolysis
fotometría, photometry
fotomicrografía, photomicrograph
fotón, photon
fotoquímica, photochemistry
fotorreceptor, photoreceptor
fotorresistencia, photo resistor
fotósfera, photosphere
fotosíntesis, photosynthesis
fototropismo, phototropism
fotovoltaica, photovoltaic
fracción, fraction
fracción decimal, decimal fraction

fracción propia, proper fraction
fracción simple, simple fraction
fraccionamiento, fractionation
fractal, fractal
fractura, fracture
frágil, brittle
fragmentación del hábitat, habitat fragmentation
fragmento, fragment
francio, francium
frecuencia, frequency
frecuencia genética, gene frequency
frenología, phrenology
frenos de aire, spoiler
frente, forehead, front
frente cálido, warm front
frente caliente, warm front
frente estacionario, stationary front
frente frío, cold front
frente ocluido, occluded front
Freón, Freon
fricción, friction, rubbing
fricción de arrastre, frictional drag
friccional, frictional
fronda, frond
frontal, frontal
frotis de sangre, blood smear
fructosa, fructose
fruta, fruit
fuente, source
fuente de agua salada, salt spring
fuente hidrotermal, thermal vent
fuente de alta temperatura, high temperature source
fuente de energía, energy source, energy resource
fuerte, high, strong, heavy
fuerza[1], force
fuerza boyante, buoyant force
fuerza centrífuga, centrifugal force
fuerza centrípeta, centripetal force
fuerza de desplazamiento del pistón, piston travel force
fuerza de elevación, uplifting forces
fuerza de fricción, force of friction
fuerza de gravedad, force of gravity
fuerza de liberación del pistón, piston release force
fuerza de sustentación, lift
fuerza de Van der Waals, Van der Waals force
fuerza débil, weak force
fuerza gravitacional, gravitational force
fuerza interna, internal force
fuerza eléctrica, electric force
fuerza electromagnética, electromagnetic force

fuerza electromotriz, electromotive force
fuerza electrostática, electrostatic force
fuerza equilibrante, equilibrant force
fuerza externa, external force
fuerza G, G-force
fuerza magnética, magnetic force
fuerza mecánica, mechanical force
fuerza normal, normal force
fuerza nuclear débil, weak nuclear force
fuerza nuclear fuerte, strong nuclear force
fuerza paralela, parallel force
fuerza perpendicular, perpendicular force
fuerza resultante, net force
fuerzas balanceadas, balanced forces
fuerzas concurrentes, concurrent forces
fuerzas de nivelación, leveling forces
fuerzas desiguales, unbalanced forces
fuerza[2], strength
fulcro, fulcrum
fumar, smoke
fumarola, fumarole
función, function
función exponencial, exponential function
función trigonométrica, trigonometric function
funcionamiento, function
funcionar, function
fundamental, fundamental
fundamento, basis, foundation
fundir[1], melt
fundir[2], smelt
fusible, fuse
fusión, fusion, meltdown, melting
fusión en caliente, hot-melt
fusión nuclear, nuclear fusion

gabro, gabbro
gadolinio, gadolinium
gafas de protección, goggles
galactosa, galactose
galactosemia, galactosemia
galaxia, galaxy
galaxias espirales, spiral galaxies
galaxias elípticas, elliptical galaxies
galaxias irregulares, irregular galaxies
galena, galena
galio, gallium
galón, gallon
galvánico, galvanic
galvanización, galvanization
galvanómetro, galvanometer
galvanoplastia, electroplating

gama globulina, gamma globulin
gameto, reproductive cell, gamete
  gameto femenino, female gamete
gametofito, gametophyte
gametogénesis, gametogenesis
gametos, sex cells
ganado, livestock
ganar, gain
ganglio, ganglion
  ganglio linfático, lymph node
  ganglio linfático, lymph gland
garganta, throat, gullet
garnate, garnet
garra[1]**,** claw
garra[2]**,** talon
garrapata, tick
gas, gas
  gas comprimido, compressed gas
  gas de agua, water gas
  gas de efecto invernadero, greenhouse gas
  gas de síntesis, synthesis gas
  gas diatómico, diatomic gas
  gas ideal, ideal gas
  gas natural, natural gas
  gas noble, noble gas
  gas pobre, producer gas
  gas raro, rare gas
  gas sanguíneo, blood gas
gasificación, gasification
gasolina, gasoline
gasterópodo, gastropod
gasto, stream flow
gástrico, gastric
gástrula, gastrula
gastrulación, gastrulation
gata peluda, gypsy moth
gatillo, trigger
gatito, kitten
gato, cat
gaviota, sea gull
géiser, geyser
gel, gel
  gel de sílice, silica gel
gelatina, gelatin
gema, gem
gemelo idéntico, identical twin
gemelos siameses, Siamese twins
gen, gene
  gen alterado, altered gene
  gen defectuoso, defective gene
  gen dominante, dominant gene
  gen recesivo, recessive gene
  gen vinculado al sexo, sex-linked gene
generación, generation
  generación de esporofitos, sporophyte generation
  generación parental, parent generation
  generación sexual, sexual generation

generador, generator
generador de corriente alterna, alternating current generator
generador eléctrico, electric generator
generalización, generalization
generar, generate
género, genus
genética, genetics
genética de poblaciones, population genetics
genéticamente, genetically
genético, genetic
genitales, genitals
genoma, genome
genotipo, genotype
genotipo heterocigótico, heterozygous genotype
geocéntrico, geocentric
geoda, geode
geofísica, geophysics
geografía, geography
geología, geology
geólogo, geologist
geomagnetismo, geomagnetism
geometría, geometry
geometría analítica, analytic geometry
geometría plana, plane geometry
geométrico, geometric
geoquímica, geochemistry
geosinclinal, geosyncline
geotérmica, geothermal
gerente de campo, field manager
germanio, germanium
germen, germ
germinación, germination
germinar, germinate
gestación, gestation
gestión de los recursos, resources management
giberelina, gibberellin
gibosa (luna), gibbous
giga, giga
gigabyte, gigabyte
gigante roja, red giant
gigantismo, giantism
gimnospermas, gymnosperm
ginecología, gynecology
ginkgo, ginkgo
girar, revolve, spin
giro, twirl
giroscopio, gyroscope
glaciación, ice age
glaciar, glacier
glaciar alpino, alpine glacier
glaciar continental, continental glacier
glándula, gland
glándula de Cowper, Cowper's gland
glándula de secreción interna, ductless gland
glándula endocrina, endocrine gland
glándula exocrina, exocrine gland
glándula mamaria, mammary gland
glándula paratiroides, parathyroid gland
glándula pineal, pineal gland, pineal body
glándula pituitaria, pituitary gland
glándula prostática, prostate gland
glándula salival, salivary gland

glándula sebácea, sebaceous gland
glándula sudorípara, sweat gland
glándula suprarrenal, adrenal gland
glaucoma, glaucoma
glicemia, blood sugar
glicérido, glyceride
glicerina, glycerin
glicerol, glycerol
glicerol gastrovascular, gastrovascular glycerol
glicina, glycine
glicol, glycol
global, global
globulina, globulin
glóbulo, blood cell
glóbulo blanco, white corpuscle
glóbulo rojo (hematíes), red corpuscle (red blood cell)
glomérulo aftosa del ganado, glomerulus
glosopeda, foot-and-mouth disease
glucagón, glucagon
glucógeno, glycogen
glucólisis, glycolysis
glucosa[1], glucose
glucosa[2], dextrose
gluones, gluon
glutamato monosódico, monosodium glutamate
glutamina, glutamine
gneis, gneiss
gobernar, govern
golfo, gulf
golondrina, swallow
golpe, strike
golpear, batter, club
gónada, gonad
gonadotropina, gonadotropin
gonorrea, gonorrhea
gordo, fat
gorila, gorilla
gota[1], drop
gota de rocío, dewdrop
gota[2], gout
gotero, dropper
gotita, droplet
grabado en acero, Steel-engraving
grabador de sonido, sound recorder
grabar, record
gradiente, gradient
gradiente adiabático seco, dry adiabatic lapse rate
gradiente de presión, pressure gradient
gradilla, test tube rack
grado, degree, rank
grado Celsius (°C), Celsius (°C)
grado de saturación, degree of saturation
grado Fahrenheit (°F), Fahrenheit (°F)
gradual, gradual
gradualismo, gradualism
gradualmente, gradually
gráfica de pastel, pie graph
gráfico[1], graph
gráfico de barras, bar graph
gráfico lineal, line graph
gráfico[2], graphic, graphical
gráfico[3], chart
gráficos, graphics
grafito, graphite

gramo (g), gram (g)
gramo de masa atómica, gram atomic mass
gramo de masa equivalente, gram equivalent mass
gramo de masa molecular, gram molecular mass
(la) gran explosión, Big Bang
grana, grana
granate, gamet
granito, granite
granizo, hail
grano[1], grain
grano de polen, pollen grain
granos de carbono, carbon grain
grano[2] (semilla), kernel
granulación, prill
grasa[1], fat
grasa de ballena, blubber
grasa no saturada, unsaturated fat
grasa saturada, saturated fat
grasa[2], grease
grava, gravel
gravedad, gravity
gravedad cero, zero gravity
gravedad específica, specific gravity
gravedad ingravidez, zero gravity
gravímetro, gravimeter
gravitación, graviton
gravitatorio, gravitational
gray (Gy), gray (Gy)
grieta[1], seam
grieta[2] (geología), cleavage, crevasse
grillo, cricket
grosor, thickness
grueso[1], fat
grueso[2], coarse
grupo, group, set
grupo amino, amino group
grupo carboxilo, carboxyl group
grupo cero, zero group
grupo círculo, set
grupo de cereales, cereal group
grupo de la tabla periódica, zero group
grupo funcional, functional group
grupo hidroxilo, hydroxyl group
grupo sanguíneo, blood group, blood type
guanina, guanine
guepardo, cheetah
guerra nuclear, nuclear war
guía de metal, metal runner
guijarro, pebble
gusano, worm
gusano de la seda, silkworm
gusanos planos, flatworm
guyot, guyot

hábitat, habitat
hábito, habit
  hábito saludable, healthy habit
  hábitos de trabajo efectivos, effective work habits
hafnio, hafnium
halcón[1]**,** falcon
halcón[2]**,** hawk
hálito, halite
halo[1]**,** corona
halo[2]**,** halo
halógeno, halogen
hardware, hardware
harina de roca, rock flour
hasta, till
haz vascular, vascular bundle
hazaña, feat
hebra, strand
heces, feces
hecho, fact
hectómetro, hectometer
helada, frost
helecho, fern
hélice[1]**,** helix
hélice[2]**,** propeller
helicóptero, helicopter
helio, helium
heliocéntrico, heliocentric
heliotropismo, heliotropism
hematita, hematite
hematoma, hematoma
hembra, female
hemiplejia, hemiplegia
hemisferio, hemisphere
  hemisferio norte, Northern Hemisphere
  hemisferio occidental, Western Hemisphere
  hemisferio oriental, Eastern Hemisphere
  hemisferio sur, Southern Hemisphere
hemofilia, hemophilia
hemoglobina, hemoglobin
hemólisis, hemolysis
hemorragia, hemorrhage
heno, hay
heparina, heparin
hepático, hepatic
hepatitis, hepatitis
heptágono, heptagon
herbicida, herbicide, weed killer
herbívoro, herbivore
heredable, heritable
heredar, inherit
hereditario, hereditary
herencia[1]**,** heredity
herencia[2]**,** inheritance
  herencia ligada al sexo, sex-linkage inheritance
  herencia mezclada, blending inheritance
  herencia poligénica, multiple-gene inheritance
hermafrodita, hermaphrodite
hermano, sibling
  hermano gemelo, fraternal twin
hernia, hernia
herpes, herpes
herpetología, herpetology
herramienta, tool
hertz, hertz
hervir, boil
hervor, boiling
heterocíclico, heterocyclic

heterocigoto, heterozygous
heterótrofo, heterotrophic
hexagonal, hexagonal
hexágono, hexagon
hibernación, hibernation
hibridación, hybridization
híbrido, hybrid
hidra, hydra
hidratación, hydration
hidratado, hydrous
hidrato, hydrate
hidráulica, hydraulics
hidráulico, hydraulic
hidrocarburo, hydrocarbon
  hidrocarburos alicíclicos, alicyclic hydrocarbons
  hidrocarburos alifáticos, aliphatic hydrocarbons
hidrocoloide, hydrocolloid
hidrodinámica, hydrodynamics
hidrofílico, hydrophilic
hidrofóbico, hydrophobic
hidrogenación, hydrogenation
hidrógeno, hydrogen
  hidrógeno pesado, heavy hydrogen
hidrogenolisis, hydrogenolysis
hidrólisis, hydrolysis
hidrología, hydrology
hidrómetro, hydrometer
hidroponía, hydroponics
hidrósfera, hydrosphere
hidrostática, hydrostatics
hidrotermal, hydrothermal
hidrotropismo, hydrotropism
hidróxido de calcio, calcium hydroxide
hidróxido de potasio, potassium hydroxide
hidróxido de sodio, sodium hydroxide
hidroxilo, hydroxyl
hidruro, hydride
  hidruro de boro, boron hydride
hielo, ice
  hielo a la deriva, pack ice
  hielo seco, dry ice
hiena, hyena
hierba[1], grass
hierba[2], herb
hierro (Fe), iron (Fe)
hígado, liver
higrofita, hygrophyte
higrómetro, hygrometer
higroscópico, hygroscopic
hilio, hilum
hilo, thread
himen, hymen
hiperactividad, hyperactivity
hipérbola, hyperbola
hipermétrope, farsighted
hipermetropía, farsightedness
hiperparasitismo, hyperparasitism
hipertensión, hypertension
  hipertensión arterial, high blood pressure
hipertexto, hypertext
hipertiroidismo, hyperthyroidism
hipervínculo, hyperlink
hipnosis, hypnosis
hipo, hiccup
hipoclorito de calcio, calcium hypochlorite
hipocondría, hypochondria
hipocotilo, hypocotyl

hipoglucemia, hypoglycemia
hipopótamo, hippopotamus
hiposecreción, hyposecretion
hipotálamo, hypothalamus
hipotenusa, hypotenuse
hipotermia, hypothermia
hipótesis, hypothesis
 hipótesis de la cerradura-llave, lock-and-key hypothesis
 hipótesis de un gen—un polipéptido, one gene-one polypeptide hypothesis
 hipótesis heterótrofa, heterotroph hypothesis
hipotético, hypothetical
hipotiroidismo, hypothyroidism
hirviente, boiling
histamina, histamine
histidina, histidine
histología, histology
histona, histone
hocico, snout
hockey, hockey
hoja, leaf
 hoja abigarrada, variegated leaf
 hoja compuesta, compound leaf
 hoja de recolección de datos, data collection sheet
hollín[1], soot
hollín[2], smut
holmio, holmium
holografía, holography
holograma, hologram
hombro, shoulder
homeopatía, homeopathy
homeostasis, homeostasis
homínido, hominid
Homo sapiens, Homo sapiens
homocíclico, homocyclic
homocigoto, homozygous
homogeneización, homogenization
homólogo, homologous
homopolímero, homopolymer
hongo[1], fungus
hongo[2], mushroom
hongos, fungi
hora estándar, standard time
horario universal, universal time
horizontal, horizontal
horizontalidad original, original horizontality
horizontalmente, horizontally
horizonte, horizon
 horizontes del suelo, soil horizon
hormiga, ant
 hormiga obrera, soldier ant, worker ant
hormigón, concrete
hormona, hormone
 hormona adrenocorticotrópica, adrenocorticotropic hormone (ACTH)
 hormona de crecimiento, growth hormone (GH)
 hormona estimulante de la tiroides, thyroid-stimulating hormone (TSH)
 hormona folículo estimulante (FSH), follicle stimulating hormone (FSH)

hormona juvenil, juvenile hormone
hormona luteinizante (HL), luteinizing hormone (LH)
hormona paratiroidea, parathyroid hormone
hormona sexual, sex hormone
hornablenda, hornblende
horno, furnace
horología, horology
horticultura, horticulture
hoy en día, present-day
huella de carbono, carbon footprint
huellas de ADN, DNA fingerprinting
huerta solar, solar farm
hueso[1], bone
hueso coxal, hipbone
hueso esponjoso, spongy bone
hueso frontal, frontal bone
hueso malar, cheek bone
hueso sólido, solid bone
hueso[2], stone (fruit)
huésped, host
hueva, roe, spawn
huevo, egg
hule, rubber
hulla, bituminous coal
humano, human
humectante, humectant
humedad, humidity, moisture
humedad absoluta, absolute humidity
humedad relativa, relative humidity
húmedo, humid
húmero, humerus
humo, fume, smoke
humor vítreo, vitreous humor
humus, humus
hundibles, sinkable
hundimiento[1], sag
hundimiento[2], subsidence
hundir, swamp
huracán, hurricane
huso, spindle

I

I+D (investigación y desarrollo), R&D (research and development)
ibuprofeno, ibuprofen
iceberg, iceberg
ictericia, jaundice
ictiología, ichthyology
ictiosauro, ichthyosaur
ideal, ideal
idéntico, identical
identidad, identity
identificación, label
identificación genética, genetic fingerprinting
identificar[1], identify
identificar los patrones, identify patterns
identificar[2], label
ígneas extrusivas, extrusive igneous
ígneo, igneous
ignición espontánea, spontaneous ignition
igual, equal

igualdad, evenness
igualdad de cambio, equality of change
igualmente, evenly
iguana, iguana
íleon, ileum
iluminación[1], illumination
iluminación[2], lighting
iluminancia, illuminance
iluminar, illuminate
ilusión, illusion
ilustrar, illustrate
imagen, image
imagen real, real image
imagen virtual, virtual image
imaginar, imagine
imaginario, imaginary
imago, imago
imán, magnet
imán permanente, permanent magnet
imán temporal, temporary magnet
imán temporario, temporary magnet
imantar, magnetize
impactar, impact
impacto de los sistemas de información, impact of information systems
impacto ambiental, environmental impact
impactos económicos, economic impacts
impartir, impart
impedancia, impedance
impermeable, waterproof
impétigo, impetigo
implantación, implantation
implicancia, implication
implicancias ambientales, environmental implications
importación, importation
importancia, significance
importar, import
imposible, impossible
impregnación, impregnation
impresora, printer (computer)
imprimir, imprint
impulsar, propel
impulso, impulse
impulso nervioso, nerve impulse
impureza, impurity
in vitro, in vitro
inactividad, dormancy
inactivo, dormant
inafectado, unaffected
inapropiado, inappropriate
incandescente, incandescent
incertidumbre, uncertainty
incidencia, incidence
incidente, incident
incipiente, budding
incisión, incision
incisivo, incisor
inclinación, bob
inclinado, inclined
inclinar, tilt
incluir, include
inclusion, inclusion
incompresible, incompressible
inconformidad, unconformity
incremento, increment
incubación, hatching
incubar, incubate
indagación científica, scientific inquiry
Indanthrene Azul, Indan-

threne Blue
independencia, independence
indicación, indication
indicador, indicator
indicador de pH, pH indicator
indicador de bicarbonato, bicarbonate indicator
indicar, indicate
índice[1], rate
índice[2], index
índice de refracción, index of refraction
índice UV, UV index
indiferencia, disregard
indígena, indigenous
indio (elemento químico), indium
indirecto, indirect
individual, individual
individuo, individual
inducción, induction
inducción electromagnética, electromagnetic induction
inducción magnética, magnetic induction
inductivo, inductive
industrialización, industrialization
inercia, inertia
inerte, inert
inestable, unstable
inestable, unsteady
infección, infection
infección bacteriana, bacterial infection
infeccioso, infectious
inferencia, inference
inferir, infer
infiltración, infiltration
infinito[1], infinity
infinito[2], infinite
inflamable, inflammable
inflamación, inflammation
inflar, inflate
inflexible, inelastic
inflorescencia, inflorescence
influenza, influenza
información, information
información sobre el consumo de productos, consumer product data
información electrónica, electronic information
infrarrojo, infrared
infrasonido, infrasound
ingeniería, engineering
ingeniería civil, civil engineering
ingeniería genética, genetic engineering
ingeniería mecánica, mechanical engineering
ingeniería química, chemical engineering
ingeniero, engineer
ingeniero de programas, software engineer
ingesta, intake
ingestión, ingestion
ingle, groins
ingrediente, ingredient
inhalación, inhalation
inhibición, inhibition
inhibidor, inhibitor
inicial, initial
iniciar, launch
injerto, graft, grafting
injerto de piel, skin graft
inmersión, immersion
inmigración, immigration

inmiscible, immiscible
inmune, immune
inmunidad, immunity
inmunidad activa, active immunity
inmunidad adquirida, acquired immunity
inmunidad mediada por células, cell-mediated immunity
inmunidad humoral, humoral immunity
inmunidad innata, inborn immunity
inmunidad materna, maternal immunity
inmunidad natural, natural immunity
inmunidad pasiva, passive immunity
inmunización, immunization
inmunodeficiencia, immunodeficiency
inmunogenicidad, immunogenicity
inmunología, immunology
inoculación, inoculation
inodoro de composta, composting toilet
inofensivo, harmless
inorgánico, inorganic
insaturación, unsaturation
insecticida, insecticide
insectívoro, insectivore
insecto, insect
insertar, insert
insolación, insolation
insolación incidente, incident insolation
insoluble, insoluble
inspeccionar, inspect
inspiración, inspiration
instalación, setup
instantáneamente, instantaneously
instantáneo, instantaneous
instante, instant
instinto, instinct
instrumento, instrument
insulina, insulin
integración, integration
integral, integral
integrar, embed, integrate
integumento, integument
inteligencia artificial, artificial intelligence
intensidad, intensity
intensidad de la insolación, intensity of insolation
intensidad de la radiación, intensity of radiation
intensidad del campo eléctrico, electric field intensity
intensidad luminosa, luminous intensity
interacción, interaction
interactuar, interact
intercambiar, switch
intercambio de gases, gas exchange
intercambio gaseoso, gaseous exchange
intercelular, intercellular
interceptar, intercept
intercrecimiento, intergrowth
interestelar, interstellar
interfase[1], interphase
interfase[2], interface
interferencia, interference
interferencia destructiva, destructive interference

interferir, interfere
interferón, interferon
interior, interior
intermedio, intermediate
internamente, internally
Internet, Internet
interneurona, interneuron
interno, internal, inward
interpartícula, interparticle
interpolar, interpolate
interpretación, interpretation
interpretar, interpret
  interpretar los datos, interpret data
interrelación, interrelationship
interrelacionar, interrelate
interrumpir, disrupt
interrupción, hiatus
interruptor, switch
intersección, intersection
intersticial, interstitial
intervalo[1], interval
  intervalo de curvas de nivel, contour interval
intervalo[2], range
intestino[1], bowel
intestino[2], gut, intestine
  intestino delgado, small intestine
  intestino grueso, large intestine
intoxicación alimentaria, food poisoning
intravenoso, intravenous
introducir, introduce
intrón, intron
intrusión, intrusion
inundación, flood
invasión, invasion
inventar, invent
inversión, inversion
  inversión de temperatura, temperature inversion
inverso, inverse
invertebrado, invertebrate
invertir[1], subvert
invertir[2], invert
investigación[1], research
investigación[2], investigation
  investigación bibliotecaria, library investigation
  investigación científica, scientific investigation
  investigación sobre los fenómenos, inquiry into phenomena
investigador, researcher
investigar, explore
invierno, winter
involucrar, involve
involuntario, involuntary
ion, ion
  ion hidronio, hydronium ion
  ion complejo, complex ion
  iones de sodio, sodium ion
ionización, ionization
ionizar, ionize
ionógeno, ionogen
ionósfera, ionosphere
iridio, iridium
iris, iris
irradiar, irradiate
irregular, irregular
irreversible, irreversible
irrigabilidad, irritability
irrigar, irrigate
isla, island
  isla barrada, barrier island

islote, islet
islotes de Langerhans, islets of Langerhans
isobara, isobar
isobara de adsorción, adsorption isobar
isocoro de van't Hoff, van't Hoff isochore
isomerización, isomerization
isómero, isomer
isómero geométrico, geometric isomer
isostasia, isostasy
isosuperficie, isosurface
isotáctico, isotactic
isoterma de van't Hoff, van't Hoff isotherm
isotermas de adsorción, adsorption isotherm
isotermo, isotherm
isótopo, isotope
istmo, land bridge
iterbio, ytterbium
itrio, yttrium

J

jabón, soap
jade, jade
jalado del agua por transpiración, transpiration pull
jaspe, jasper
jeringa, syringe
joule, joule
joyero, jeweler
juego, set
jugo gástrico, gastric juice
jugo intestinal, intestinal juice
jugo digestivo, digestive juice
jugo pancreático, pancreatic juice
julio, joule
jungla, jungle
junta, gasket
juntar, collect
Júpiter, Jupiter
juvenil, juvenile
juventud, youth

kame, kame
Kelvin, Kelvin
kilo, kilo
kilobit, kilobit
kilobyte, kilobyte
kilogramo (kg), kilogram (kg)
kilohertz, kilohertz
kilolitro (kl), kiloliter (kl)
kilómetro (km), kilometer (km)
kilopascal, kilopascal
kilovatio, kilowatt
kilovatio hora, kilowatt hour
koala, koala
krill, krill
kriptón, krypton

L

laberinto, labyrinth
labio, lip
laboratorio, laboratory
lacolito, laccolith
lactancia, lactation
lactasa, lactase
Láctea, Lactea
lácteo, lacteal

lactona, lactone
lactosa, lactose
ladera, hillslope
ladrillo, brick
ladrón, burglar
lagarta peluda, gypsy moth
lagarto, lizard
lago, lake
  lago de la hoya glaciar, kettle lake
  lagos glaciares, glacial lakes
  Lagos Finger, Finger Lakes
lágrima, tear
laguna, lagoon
Lamarckismo, Lamarckism
lámina, lamella
laminaria (alga), kelp
lámpara fluorescente, fluorescent lamp
lámpara incandescente, incandescent lamp
lampírido, firefly
lamprea, lamprey
lanceta, lancet
Langerhans, Langerhans
langosta, lobster
lanolina, lanolin
lantánido, lanthanide
lantano, lanthanum
llanura fluvio glacial, outwash plain
llanura de aluvión, outwash plain
lanzacohetes, rocket launcher
lanzador de peso, shot-putter
lanzar[1], cast
lanzar[2], launch
laringe, larynx
laringitis, laryngitis
larva, larva
láser, laser
lateral, sideway
látex, latex
latido, beat
latitud, latitude
latón, brass
laurencio, lawrencium
lava, lava
laxante, laxative
lecha seminal de los peces, milt
leche, milk
lecho, bedding, bed
  lecho del río, stream bed
  lecho rocoso, bedrock
lechuza, owl
lecitina, lecithin
LED (diodo emisor de luz), LED (light-emitting diode)
legumbre, legume
lejano, remote
lejía, lye
lémur, lemur
lengua, tongue
lengüeta, barb
lengüeta de tiro, pull-tab
leño, log
lente, lens
  lente compuesta, compound lens
  lente cóncava, concave lens
  lentes convexa, convex lens
  lente divergente, diverging lens
  lente objetivo, objective lens
lenteja de agua, duckweed
león marino, sea lion

leopardo, leopard
lepra, leprosy
leptón, lepton
leptospiraceae, spirochete
lesión por congelación, frostbite
leucemia, leukemia
leucina, leucine
leucocito, white blood cell, leukocyte
levadura, yeast
  levadura en polvo, baking powder
levantar, lift
ley, law
  ley científica, scientific law
  ley combinada de gas, combined gas law
  ley de acción y reacción, law of action and reaction
  ley de conservación, conservation law
  Ley de Coulomb, Coulomb's Law
  ley de distribución de Nernst, Nernst distribution law
  ley de distribución independiente, law of independent assortment
  ley de dominio, law of dominance
  ley de elasticidad de Hooke, Hooke's Law
  ley de gravitación, law of gravitation
  Ley de Gravitación de Newton, Newton's Law of Gravitation
  ley de Hardy-Weinberg, Hardy-Weinberg Law
  Ley de Hubble, Hubble's Law
  Ley de la Composición Constante, Law of Constant Composition
  ley de la segregación, law of segregation
  Ley de Lenz, Lenz's Law
  Ley de Ohm, Ohm Law
  Ley de Reflexión, Law of Reflection
  ley de reparto, Nernst distribution law
  Ley de Snell, Snell's Law
  ley del gas ideal, ideal gas law
  ley del inverso del cuadrado, inverse-square law
  ley del uso y desuso, law of use and disuse
  ley periódica, periodic law
  Leyes de Movimiento de Newton, Newton's laws of motion
Leyden Jar, Leyden Jar
leyenda, legend
libélula, dragonfly
liberar[1], release
liberar[2], liberate
libertad, freedom
libra, pound
libreta, booklet
LIC (líquido intracelular), ICF (intercellular fluid)
licor, liquor
licuar, liquefy

licuefacción, liquefaction
liebre, hare
ligación genética, gene linkage
ligamento, band, ligament
ligar, bind
ligeramente, slightly
lignina, lignin
lignito, lignite
lima, lime
limbo, blade
limitación de nutrientes, limiting nutrient
limitaciones de los sistemas de información, limitations of information systems
límite[1], boundary
  límite convergente, convergent boundary
  límite de colisión, collision boundary
  límite de placas convergentes, convergent plate boundary
  límite de placas divergentes, divergent plate boundary
  límite frontal, frontal boundary
límite[2], limit
límite[3], line
  límite de las nieves perpetuas, snow line
  límite del arbolado, timberline
límite[4], threshold
limo[1], ooze
limo[2], loam
limolita, siltstone
limón, lime
limonita, limonite
linaje, pedigree
línea, line
  línea antinodal, antinodal line
  línea continental, continental line
  línea costera, coastline
  línea de campo eléctrico, electric field line
  línea de campo magnético, magnetic field line
  línea de fuerza, line of force
  línea de nieve, snow line
  línea de nivel, isoline
  línea de primer orden, first-order line
  línea de segundo orden, second-order line
  línea divisoria de aguas, watershed
  línea internacional de cambio de fecha, International Date Line
  línea lateral, lateral line
  línea nodal, nodal line
  líneas espectrales, spectral lines
lineal, linear
linfa, lymph
linfocito, lymphocyte
liofílico, lyophilic
liofilización, freeze drying
lipasa, lipase
lípido, lipid
liquen, lichen
líquido, liquid

líquido cefalorraquídeo (LCR), cerebrospinal fluid
líquido cerebroespinal (LCR), cerebrospinal fluid
líquido de frenos, brake fluid
líquido intracelular (LIC), intercellular fluid (ICF)
líquido seminal de los peces, milt
líquido viscoso, viscous liquid
líquido volátil, volatile liquid
lirio, lily
lista, list
litargirio, litharge
literalmente, literally
litio, lithium
litología, petrology
litoral, littoral
litósfera, lithosphere
litro (L), liter (L)
lixiviación, leach
llama, flame
llamarada, flare
llanura, plain
llanura abisal, abyssal plains
llanura aluvial, flood plain
llave maestra, skeleton key
llevar a cabo, accomplish
llevar la energía, carrying power
lluvia, rain
lluvia ácida, acid rain
lo menor, least
lóbulo, lobe
lóbulo frontal, frontal lobe
lóbulo occipital, occipital lobe
lóbulo olfativo, olfactory lobe
lóbulo parietal, parietal lobe
lóbulo temporal, temporal lobe
locomoción, locomotion
lodo[1], mud
lodo[2], sludge
lodo mineral, slime
loess, loess
logaritmo, logarithm
logaritmo común, common logarithm
logaritmo natural, natural logarithm
lógica, logic
lograr, achieve
lombriz, earthworm, worm
lomo, back
longitud[1], length
longitud de onda, wavelength
longitud de una sombra, length of a shadow
longitud[2], longitude
loro, parrot
low pressure area, borrasca, zona de baja presión
luciérnaga, firefly
lubina, sea bass
lubricante, lubricant
lugar, location
lugar decimal, decimal place

lumbar, lumbar
lumen, lumen
luminiscente, luminescent
luminoso, luminous
luna, moon
  luna creciente, crescent moon
  luna llena, full moon
  luna nueva, new moon
lunar, lunar
lupa, magnifying glass
  lupa de mano, hand lens
lupia, burl
lustre, luster
lutecio, lutetium
luz, light
  luz estroboscópica, strobe light
  luz relámpago, photoflash
  luz visible, visible light
lycophyta, club moss

maceta, pot
macho, male
macroanálisis, macroanalysis
macroclima, macroclimate
macronúcleo, macronucleus
mácula, macula
madera, wood
  madera roja, redwood
madre biológica, natural mother
madriguera, warren
maduración, maturation
madurez, maturity
  madurez sexual, sexual maturity
máficas, mafic
magenta, magenta
magma, magma
magnalium, magnalium
magnesia, magnesia
magnesio, magnesium
magnesita, magnesite
magnético, magnetic
magnetismo, magnetism
magnetita, lodestone, magnetite
magnetizar, magnetize
magnetómetro, magnetometer
magneton, magneson
magnetoquímica, magnetochemistry
magnetrón, magnetron
magnificación biológica, biological magnification
magnificador, magnifier
magnitud, magnitude
  magnitud absoluta, absolute magnitude
  magnitud aparente, apparent magnitude
  magnitud del terremoto, earthquake magnitude
mal de Alzheimer, Alzheimer's disease
mal funcionamiento, malfunction
malaquita, malachite
  malaquita azul, azurite
malaria, malaria
malatión, malathion

maleabilidad, malleability
maleza[1]**,** undergrowth
maleza[2]**,** weed
maligno, malignant
maloclusión, malocclusion
maltasa, maltase
maltosa, maltose
mamífero, mammal
  mamífero marsupial, pouched mammal
  mamífero placentario, placental mammal
  mamífero sin placenta, nonplacental mammal
mamografía, mammogram
mamut, mammoth
  mamut lanudo, woolly mammoth
manatí, manatee
mancha, stain
  mancha ocular, eyespot
  mancha solar, sunspot
mandíbula, jaw
manejar, manage, handle
manganeso, manganese
manglar, mangrove
mangle, mangrove
mangosta, mongoose
manguito rotador, rotator cuff
manipular, handle
manitol, mannitol
mano, hand
manómetro, manometer
mantener, maintain
mantenimiento, maintenance
mantisa, mantissa
manto, mantle
  manto glaciar, icecap
mapa batimétrico, bathymetric map
mapa de bits, bitmap
mapa de contorno, contour map
mapa de isolíneas, isoline map
mapa genético, genetic map
mapa meteorológico sinóptico, synoptic weather map
máquina, machine
  máquina corazón-pulmón, heart-lung machine
  máquina de vapor, steam engine
  máquina simple, simple machine
mar, sea
marcador genético, genetic marker
marcapasos, pacemaker
marchitarse, wilt
marco, framework
  marco de referencia, frame of reference
marea, tide
  marea alta, high tide
  marea baja, low tide
  marea creciente, flood tide
  marea roja, red tide
  marea viva, spring tide
  marea viva y muerta, neap tide
  mareas de tempestad, storm surge

marejada, tidal wave
maremoto, tidal wave
marfil, ivory
margen, margin
marinero, sailor
marino, marine
mariposa, butterfly
mariquita, ladybug
marisma, marsh
mármol, marble
marsopa, porpoise
marsupial, marsupial
Marte, Mars
martillo, hammer
más allá, beyond
más grueso, thicker
más importante, greatest
masa, mass
  masa atómica, atomic mass
  masa continental, landmass
masa crítica, critical mass
  masa de aire, air mass
  masa de aire polar marítima, maritime polar airmass
  masa de aire tropical marítima, maritime tropical airmass
  masa gravitacional, gravitational mass
  masa inerte, inertial mass
  masa molecular, molecular mass
  masa terrestre, land mass
  masa de aire tropical continental, continental tropical air mass
  masa equivalente, equivalent mass
máser, maser
masilla, putty
masivo, massive
masticación, mastication
mastodonte, mastodon
mastoideo, mastoid
matemáticamente, mathematically
matemáticas, mathematics
matemático, mathematician
materia, matter
  materia gris, gray matter
  materia prima, feedstock
material, material
  material genético, genetic material
  material inflamable, flammable material
  material radiactivo, radioactive material
  materiales electrónicos, electronic materials
matices, overtones
matiz, hue
matraz, flask
  matraz aforado, volumetric flask
  matraz de Erlenmeyer, Erlenmeyer flask
  matraz de Florencia, Florencia flask
matriz[1]**,** array
matriz[2]**,** matrix
matriz[3]**,** womb
maxilar, jaw
  maxilar inferior, mandible
  maxilar mandíbula, mandible

máximo, maximum
mayor, greatest
meandro, oxbow lake
mecánica, mechanics
 mecánica clásica, classical mechanics
 mecánica cuántica, quantum mechanics
 mecánica ondulatoria, wave mechanics
mecánico, mechanical
mecanismo, mechanism
 mecanismo de reacción, reaction mechanism
 mecanismo de retroalimentación, feedback mechanism
mechero de Bunsen, Bunsen burner
media[1], mean
 media aritmética, arithmetic mean
media[2], half
media luna, crescent
mediana, median
medicina, medicine
 medicina alternativa, alternative medicine
 medicina forense, forensic science
 medicina interna, internal medicine
 medicina veterinaria, veterinary medicine
medición, measurement
médico, medical
medida, measure
 medida para áridos, dry measure
medio, medium
 medio ambiente, environment
mediodía, high noon
 mediodía local, local noon
 mediodía solar, solar noon
medios de comunicación, media
medir, measure
médula[1], pith
médula[2], marrow
 médula espinal, spinal cord
 médula ósea, bone marrow
 médula suprarrenal, adrenal medulla
medusa, jellyfish
mega, mega
megabit, megabit
megabyte, megabyte
megahertz, megahertz
megavatio, megawatt
megavoltio, megavolt
meiosis, meiosis
mejilla, cheek
mejillón, mussel
 mejillón cebra, zebra mussel
mejor, best
melanina, melanin
melanismo industrial, industrial melanism
melanoma, melanoma
melodía, tune
membrana, membrane
 membrana celular, cell membrane
 membrana mucosa, mucous membrane
 membrana nuclear, nuclear membrane (envelope)
 membrana permeable

selectiva, selectively permeable membrane
membrana plasmática, plasma membrane
membrana semipermeable, semipermeable membrane
membranas de la placenta, placental membranes
membranas embrionarias, embryonic membrane
membranoso, membranous
memoria, memory
memoria de solo lectura (ROM), read-only memory (ROM)
mena, ore
menarquía, menarche
mencionar, mention
mendelevio, mendelevium
Mendelismo, Mendelism
menguante, waning
meningitis, meningitis
menisco, meniscus
menopausia, menopause
menos, fewer
menstruación[1], menses
menstruación[2], menstruation
mensurable, measurable
mentol, menthol
mentón, chin
mercurio[1], mercury
Mercurio[2], Mercury
meridiano, meridian
mes, month
mes sideral, sideral month
mes sinódico, synodic month
mesa, table
mesencéfalo, midbrain
mesenterio, mesentery
meseta[1], mesa
meseta de escasa pluviosidad de Sudáfrica, veldt
meseta[2], plateau
meseta de lava, lava plateau
mesodermo, mesoderm
mesófilo, mesophyll
mesófilo en empalizada, palisade mesophyll
mesófilo esponjoso, spongy mesophyll
mesófito, mesophyte
mesón, meson
mesopausa, mesopause
mesósfera, mesosphere
meta no definida, undefined target
metabólico, metabolic
metabolismo, metabolism
metacarpiano, metacarpal
metafase, metaphase
metal, metal
metal alcalino, alkali metal
metal de tierra alcalina, alkaline earth metal
metal ligero, light metal
metal pesado, heavy metal
metal raro, rare metal
metálico, metallic
metaloide, metalloid
metalurgia, metallurgy
metalurgia de polvos, powder metallurgy
metamórfico, metamorphic

metamorfismo, metamorphism
metamorfismo de contacto, contact metamorphism
metamorfosis, metamorphosis
metanal, formaldehyde
metano, methane
metanol, methanol
metástasis, metastasis
metazoario, metazoan
meteorito, meteorite
meteoro, meteor
meteoroide, meteoroid
meteorología, meteorology
meticulosamente, meticulously
metilamina, methylamine
metilo, methyl
método, method
método de datación por radiocarbono, radiocarbon method
método sistemático, systematic method
métrico, metric
metro (m), meter (m)
metro cúbico, cubic meter
mezcla, mixture, compounding
mezclado, mixed
mezclar, mix
mica, mica
micelio, mycelium
micelio secundario, secondary mycelium
micelios, mycelia
micología, mycology
micro-, micro-
microanálisis, microanalysis
microbio, microbe
microbiología, microbiology
microclima, microclimate
microclina, microcline
microdisección, microdissection
microfaradios, microfarad
microfilamentos, microfilament
micrófono, microphone
micrografía, micrograph
micrómetro, micrometer, micrometer caliper
micronúcleo, micronucleus
microondas, microwave
microorganismo, microorganism
microprocesador, microprocessor
microscópico, microscopic
microscopio, microscope
microscopio compuesto, compound microscope
microscopio de contraste de fase, phase contrast microscope
microscopio de disección, dissecting microscope
microscopio de luz, light microscope
microscopio electrónico, electron microscope, electronic microscope
microscopio estereoscópico, stereomicroscope
microscopio óptico, optical microscope
microscopio simple, simple microscope
microsegundo, microsecond
microtúbulo, microtubule

mielina, myelin
miembro, member
migración, migration
migraña, migraine
migratorio, migratory
milésimo, mil
mili, milli
miliamperios, milliampere
milibar, millibar
miligramo (mg), milligram (mg)
mililitro (ml), milliliter (ml)
milímetro (mm), millimeter (mm)
  milímetro de mercurio, torr
milla, mile
  milla náutica, nautical mile
  milla terrestre, statute mile
milpiés, millipede
mimetismo, mimicry
mina a cielo abierto, strip mine
mineral, mineral
  minerales formadores de roca, rock-forming minerals
mínimo, minimum
  mínimo común denominador, lowest common denominator
minuto, minute
miofibrilla, myofibril
miope, nearsighted
miopía, myopia, nearsightedness, shortsightedness
mirada, glance
miscible, miscible
mitocondrio, mitochondrion (pl. mitochondria)
mitocondrios, mitochondria
mitosis, mitosis
moa, moa
mochila, knapsack
moda, mode
modelo, model
  modelo científico, scientific model
  modelo de compartimentos, box modeling
  modelo físico, physical model
  modelo geocéntrico, geocentric model
  modelo heliocéntrico, heliocentric model
  modelo nuclear, nuclear model
  modelo quark de los nucleones, quark model nucleon
módem, modem
moderador, moderator
modificación, modification
modificado, altered
modificar, modify, alter
modo, manner
modular, modulate
módulo, module
  módulo lunar, lunar module
Moho[1], Moho
moho[2], mold
  moho mucilaginoso, slime mold
moho[3], mildew
molalidad, molality
molécula, molecule
  molécula covalente, covalent molecule
  molécula no polar, nonpolar molecule

molécula polar, polar molecule
molécula receptora, receptor molecule
molécula transportada por el aire, airborne molecule
molecular, molecular
molibdneo, molybdenum
molino de viento, windmill
molleja, gizzard
molusco, mollusk
moluscos, mollusca
momento[1], time
momento culminante, high point
momento[2], momentum
momento angular, angular momentum
momento inicial, initial momentum
momento lineal, linear momentum
momentos, momenta
mónera, Monera
monitor, monitor
mono, monkey
monocitos, monocyte
monoclínico, monoclinic
monocotiledónea, monocotyledon
monocotiledóneo, monocotyledon
monocromático, monochromatic
monocultivo, monoculture
monodáctilo, monodactyl
monoinsaturado, monounsaturated
monómero, monomer
monominerálico, monomineralic
monomio, monomial
monomolecular, monomolecular
monoploide, monoploid
monosacárido, monosaccharide
monóxido, monoxide
monóxido de carbono, carbon monoxide
montaña, mountain
montañoso, hilly
monte aislado, butte
monte submarino, seamount, guyot
monzón, monsoon
moqueta, carpet
MOR (movimiento ocular rápido), REM (Rapid Eye Movement)
morfina, morphine
morfología, morphology
morrena, moraine
morrena de recesión, recessional moraine
morrena final, end moraine
morrena lateral, lateral moraine
morrena media, medial moraine
morrena terminal, terminal moraine
morro, snout
mortal, fatal
mortalidad, mortality
mórula, morula
mosaico, mosaic
mosaico fluido, fluid mosaic
mosca, fly (biology)
mosca de la fruta, fruit fly
moscovita, muscovite
mosquito[1], mosquito

mosquito[2], gnat
mota, speck
motilidad[1], motile
motilidad[2], motility
moto de nieve, snowmobile
motor[1], engine
  motor cohete, rocket engine
  motor diesel, diesel engine
  motor térmico, heat engine
motor[2], motor
mouse, mouse (computer)
movimiento, motion, movement
  movimiento ameboide, ameboid movement
  movimiento aparente, apparent motion
  movimiento aparente de los planetas, apparent motion of the planets
  movimiento armónico simple, simple harmonic motion
  movimiento ciliar, ciliary motion
  movimiento circular uniforme, uniform circular motion
  movimiento diario aparente, apparent daily motion
  movimiento de proyectiles, projectile motion
  movimiento diario, daily motion
  movimiento ocular rápido (MOR), Rapid Eye Movement (REM)
  movimiento retrógrado, retrograde motion
  movimiento vibratorio, vibrational motion
  movimientos celestes, celestial motions
  movimientos terrestres, terrestrial motions
mucosa, mucus
muda, molt
muela, molar
  muela del juicio, wisdom tooth
muelle[1], spring
muelle[2], dock, pier
muestra, sample
  muestra atmosférica, atmospheric cross-section
  muestra en montaje húmedo, wet-mount slide
  muestras de circuitos, loop samples
multicelular, multicellular
múltiple, multiple
multiplicación, multiplication
  multiplicación vegetativa, vegetative propagation
multiplicador, multiplier
multiplicando, multiplicand
multiplicar, multiply
múltiplo, multiple
  múltiplo común, common multiple
multirresistencia, multiple resistance
mundo animal, Animalia
muñeca, wrist
muón, muon
murciélago, bat
muscular, muscular
músculo, muscle

músculo cardíaco, heart muscle, cardiac muscle
músculo esquelético, skeletal muscle
músculo estriado, striated muscle
músculo involuntario, involuntary muscle
músculo liso, smooth muscle
músculo longitudinal, longitudinal muscle
músculo visceral, smooth muscle, visceral muscle
músculo voluntario, voluntary muscle
musgo, moss
muslo, thigh
mutación[1]**,** mutation
mutación espontánea, spontaneous mutation
mutación puntual, point mutation
mutado, mutated
mutágeno, mutagen
mutante, mutant
mutar, mutate
mutualismo, mutualism

nadir, nadir
nafta, naphtha
naftalina, naphthalene
nailon, nylon
nalgas, buttocks
nano, nano
nanómetro, nanometer
nanosegundo, nanosecond
nanotubo, nanotube
narcótico, narcotic
nariz, nose
nasal, nasal
natural, natural
naturaleza, nature
naturaleza de la superficie, nature of the surface
naturalista, naturalist
naturalizar, naturalize
naturalmente, naturally
nautilo, nautilus
navaja, razor
nave espacial, spacecraft
navegador, browser
navegar, surf
nebulosa, nebula
nebulosa planetaria, planetary nebula
necesidades energéticas locales, local energy needs
néctar, nectar
nefrona, nephron
negativo, negative
negrita, boldface
negro, black
negro de carbón, carbon black
negro de carbón, impingement black
negro de marfil, ivory black
nematocisto, nematocyst
nematodo, nematode
neodimio, neodymium
neón, neon
neopreno, neoprene
neptunio, neptunium

Neptuno, Neptune
nervio, nerve
nervio auditivo, auditory nerve
nervio ciático, sciatic nerve
nervio craneal, cranial nerve
nervio espinal, spinal nerve
nervio olfativo, olfactory nerve
nervio óptico, optic nerve
nervio vago, vagus nerve
nervio vestibulococlear, auditory nerve
nervios motores, motor nerve
neumático[1]**,** pneumatic
neumático[2]**,** tire
neumonía, pneumonia
neumonía bacteriana, bacterial pneumonia
neural, neural
neurohormona (neurotransmisor), neurohormone (neurotransmitter)
neurología, neurology
neurona[1]**,** nerve cell
neurona[2]**,** neuron
neurona de asociación, association neuron
neurona motora, motor neuron
neurona sensorial, sensory neuron
neurotransmisor, neurotransmitter
neutral, neutral
neutralización equivalente, neutralization equivalent
neutrino, neutrino
neutrón, neutron
neutrón primero, first neutron
nevada, snowstorm
nevar, snow
nevasca, snowstorm
nevé (neviza), névé (firn)
nevisca, snowstorm
Newton, Newton
niacina, niacin
nicho, niche
nicho ecológico, ecological niche
nicotina, nicotine
niebla, fog
niebla tóxica, smog
nieve, snow
nimboestrato, nimbostratus
nimbostratus, nimbostratus
nimbus, nimbus
ninfa, nymph
El Niño, El Niño
niobio, niobium
níquel, nickel
nitración, nitration
nitrato, nitrate
nitrato de bario, barium nitrate
nitrato de plata, silver nitrate
nitrato de potasio, potassium nitrate
nitrato de sodio, sodium nitrate
nitrificación, nitrification
nitrilo, nitrile
nitrito, nitrite
nitro, niter
nitrogenado, nitrogenous
nitrógeno, nitrogen

nitroglicerina, nitroglycerin
nitruración, nitriding
nitruro, nitride
  nitruro de boro, boron nitride
nivel, level
  nivel de energía, energy level
  nivel del mar, sea level
  nivel inferior, base level
  nivel trófico, trophic level
no disminuido, undiminished
no electrólito, nonelectrolyte
no metálico, non-metallic, nonmetal
no renovable, nonrenewable
no saturado, unsaturated
no transmisible, noncommunicable
Nobel, Alfred B., Nobel, Alfred B.
nobelio, nobelium
nocivo, harmful
nocturno, nocturnal
nodal, nodal
nodo, node
nódulo, nodule
  nódulo sinoauricular, sino-atrial node (S-A node)
  nódulos radicales, root nodule
nogal, walnut
nombre científico, scientific name
nombre químico, chemical name
nomenclatura, nomenclature
noradrenalina, noradrenaline
norepinefrina, noradrenaline
normalidad, normality
notación, notation
  notación científica, scientific notation
  notación decimal, decimal notation
  notación exponencial, exponential notation
notocorda, notochord
nova, nova
novillo, calf
nube, cloud
  nube cirro, cirrus cloud
  nube de electrones, electron cloud
  nube estrato, stratus cloud
nuclear, nuclear
nucleido, nuclide
núcleo[1], core
  núcleo externo, outer core
  núcleo interno, inner core
núcleo[2], nucleus (pl. nuclei)
  núcleo endospermo, endosperm nucleus
núcleo[3] (átomo), kernel
nucléolo, nucleolus
nucléolos, nucleoli
nucleón, nucleon
núcleos, nuclei
  núcleos de los espermatozoides, sperm nuclei
nucleótido, nucleotide
nuez, walnut
numerador, numerator
numeral, numeral
numéricamente, numerically

número, number
número arábigo, Arabic numeral
número atómico, atomic number
número cardinal, cardinal number
número complejo, complex number
número compuesto, composite number
número cuántico, quantum number
número de masa, mass number
número de transferencia, transference number
número entero, integer, whole number
número googol, googol
número imaginario, imaginary number
número impar, odd number
número irracional, irrational number
número mixto, mixed number
número natural, natural number
número ordinal, ordinal number
número par, even number
número perfecto, perfect number
número primo, prime number
número racional, rational number
número real, real number
numerosos, numerous
nutria, otter
nutrición, nutrition
nutrición heterotrófica, heterotrophic nutrition
nutrición autótrofa, autotrophic nutrition
nutricional, nutritional
nutriente, nutrient
nutrir, nourish
nylon, nylon

oasis, oasis
obedecer, obey
objetivo, object, objective
objeto celeste, celestial object
oblicuo, oblique
oboe, oboe
observable, observable
observación, observation
observaciones reales, actual observations
observar, observe, note
observatorio, observatory
obsidiana, obsidian
obstáculo, obstacle
obstetricia, obstetrics
obtener, obtain
obtenido, obtained
obturador, stopper
ocasionalmente, occasionally
oceánica, oceanica
océano, ocean
oceanografía, oceanography
oclusión, occlusion

octaedro, octahedron
octagonal, octagonal
octágono, octagon
octano, octane
octavo, octave
octeto, octet
ocular[1], eyepiece
ocular[2], ocular
ocultación, occultation
ocupar, occupy
ocurrir, occur
odontología, dentistry
offset, offset
oficial, official
oficio, craft
oftalmología, ophthalmology
ohmio, ohm
oído, ear
  oído externo, outer ear
  oído interno, inner ear
  oído medio, middle ear
oír, hear
ojo, eye
  ojo compuesto, compound eye
ola, wave
  ola primaria, primary (P) waves
olefina, olefin
oler, smell
olfativo, olfactory
olfato[1], smell
olfato[2], olfaction
olfatorio, olfactory
olivino, olivine
olla, pot
olmo, elm
olor, odor, smell
ombligo, umbilicus
omnívoro, omnivore
oncogén, oncogene
oncología, oncology
onda[1], wave
  onda de longitud, longitudinal wave
  onda de longitudinal, longitudinal wave
  onda de luz, light wave
  onda de materia, matter wave
  onda de radio, radio wave
  onda de sonido, sound wave
  onda de superficie, surface wave
  onda electromagnética, electromagnetic wave
  onda estacionaria, standing wave
  onda estacionaria, stationary wave
  onda expansiva, shockwave
  onda incidente, incident wave
  onda longitudinal, longitudinal wave
  onda P, P-wave
  onda polarográfica, polarographic wave
  onda primaria, primary wave
  onda S, S-wave
  onda secundaria, secondary wave
  onda sísmica, seismic wave
  onda transmitida, transmitted wave

onda transversal, transverse wave
ondas de radio, airwaves
onda[2], ripple
ónice, onyx
ónix, onyx
ontogenia, ontogeny
onza, ounce (oz)
onza líquida, fl. oz. (fluid ounce)
ooquistes, oocyst
órgano sexual, sex organ
opacidad, opacity
opaco, opaque
ópalo, opal
operación, operation
operaciones conjuntas, joint ventures
opérculo, operculum
operón, operon
opinión, opinion
opio, opium
oponerse, oppose
oportunidad, opportunity
óptica, optics
óptica de rayos, ray optics
óptico, optic, optical
óptimo, optimum
oral, oral
orangután, orangutan
órbita, orbit
orbital, orbital
orbitar, orbit
orca, orca
orden, order
orden de reacción, order of reaction
orden subordinado, suborder
ordenada, ordinate
ordenar[1], order
ordenar[2], arrange
orgánico, organic
organismo, organism
organismo saprofito, decomposer
organizar, organize
órgano, organ
órgano del sentido, sense organ
órganos rudimentarios, rudimentary organ
órganos sensoriales, sensory organs
organometálica, organo-metallic
orgánulo, organelle
orientación, orientation
orientación de la superficie, surface orientation
orificio nasal[1], blowhole
orificio nasal[2], nostril
origen, origin
original, original
orilla[1], shoreline
orilla[2], shore
orina, urine
Orión, Orion
ornitología, ornithology
ornitorrinco, platypus
oro, gold
orquídea, orchid
ortoclasa, orthoclase
ortopedia, orthopedics
ortorrómbico, orthorhombic
oruga, caterpillar
Osa Mayor, Big Dipper, Ursa Major
Osa Menor, Little Dipper, Ursa Minor
oscilación[1], oscillation
oscilación[2], swing
oscilar, swing

osciloscopio, oscilloscope
osificación, ossification
osmio, osmium
ósmosis, osmosis
oso, bear
osteoartritis, osteoarthritis
osteoblastos, osteoblast
osteocitos, osteocyte
osteología, osteology
osteoporosis, osteoporosis
ostra, oyster
otoño, autumn
óvalo, oval
ovario, ovary
ovarios, ovaries
oviducto, oviduct
oviparidad, oviparity
ovíparo, oviparous
ovipositor, ovipositor
ovocito, oocyte
ovogénesis, oogenesis
ovovivíparo, ovoviviparous
ovulación, ovulation
óvulo[1], ovum (pl. ova)
óvulo[2], ovule
óvulos, ova (sing. ovum)
oxidación, oxidation
  oxidación lenta, slow oxidation
oxidante fotoquímico, photochemical oxidant
oxidar, oxidize
óxido, oxide, rust
  óxido de calcio, calcium oxide
  óxido de cinc, zinc oxide
  óxido de hierro, iron oxide
  óxido de mangánico, manganic oxide
  óxido nítrico, nitric oxide
  óxido nitroso, nitrous oxide
oxidrilo, hydroxyl
oxigenar[1], oxygenate
oxigenar[2], aerate
oxígeno, oxygen
  oxígeno ácido, oxygen acid
oxihemoglobina, oxyhemoglobin
ozono, ozone

pabellón auricular, pinna
padre sustituto, surrogate parent
página web, webpage (computer)
paisaje, landscape
  paisaje regional, landscape region
  paisaje joven, young landscape
paja, straw
pajarera, aviary
pala, shovel
paladar, palate
paladio, palladium
palanca, lever
Paleoceno, Paleocene
paleontología, paleontology
Paleozoico, Paleozoic
palillo, toothpick
palma, palm
palmeado, palmate
palo, stick
pan de molde, bread mold
páncreas, pancreas
pandemia, pandemic
Pangea, Pangaea

panícula, panicle
pantalla de cristal líquido, LCD
pantano[1], bayou, swamp
  pantano de mangle, mangrove swamp
pantano[2], slough
papada, dewlap
papel, role
  papel tornasol, litmus paper
paperas, mumps
papila, papilla
  papila gustativa, taste bud
papión, baboon
paquete, bundle
  paquete de software, software package
paquidermo, pachyderm
par, pair
  (en) pares, paired
  par conjugado, conjugate pair
  par de bases, base pair
  par de orbitales, orbital pair
  par motor, torque
par craneal, cranial nerve
parábola, parabola
parabólica, parabolic
paracaídas, parachute
paracaidista, skydiver
parafina, paraffin
paralaje, parallax
paralelismo, parallelism
paralelo, parallel
paralelogramo, parallelogram
parálisis cerebral, cerebral palsy
parálisis infantil, infantile paralysis
paramecio, paramecium
parámetro, parameter
paraplejia, paraplegia
parasitismo, parasitism
parásito, parasite, pest
parcela, plot
parche, patch
parcial, partial
parcialidad de polarización, bias voltage
parecerse, resemble
pared celular, cell wall
pared inclinada, hanging wall
parénquima, parenchyma
parpadear, blink
parpadeo, blink
párpado, eyelid
pársec, parsec
parte[1], part
  partes por millón, parts per million
parte[2], portion
partenogénesis, parthenogenesis
partícula, particle, particulate, speck
  partícula alfa, alpha particle
  partícula beta, beta particle
  partícula elemental, elementary particle
  partícula fundamental, fundamental particle
  partícula subatómica, subatomic particle
particular, particular
(a) partir de la cual, based on which
parto, labor

parto múltiple, multiple birth
parto prematuro, premature birth
pascal, pascal
pascalio, pascal
paseriforme, passerine
pasividad, passivity
paso determinante de la velocidad, rate-determining step
pasteurización, pasteurization
pata, foot
patinaje, skating
patógenico, pathogenic
patógeno[1]**,** pathogen
patógeno[2]**,** pathogenic
patología, pathology
patrón, pattern
patrón de enrejado, trellis pattern
patrón de flujo de drenaje, stream draining pattern
patrón dendrítico, dendritic pattern
patrón radial, radial pattern
patrón rectangular, rectangular pattern
patrones de drenaje, drainage patterns
patrones latitudinales de clima, latitudinal climate patterns
pavimento, pavement
pecho, chest, breast
peciolo, leafstalk
pectina, pectin
pectoral, pectoral
peculiar, peculiar
peculiaridad, peculiarity
pediatría, pediatrics
pedicelo, pedicel
pedigrí, pedigree
pedúnculo, peduncle
pedúnculo ocular, eyestalk
pegamento, glue
pegmatita, pegmatite
pelágico, pelagic
pelagra, pellagra
pelaje, hackle
pelícano, pelican
película, flick
película delgada, thin film
peligro, hazard
peligro industrial, industrial hazard
peligros geológicos, geologic hazard
peligroso, dangerous
pelo, hair
pelota de goma, rubber ball
pelvis, pelvis
pendiente[1]**,** downslope
pendiente[2]**,** slope
péndulo, pendulum
péndulo de Foucault, Foucault pendulum
pene, penis
penicilina, penicillin
penillanura, peneplane
península, peninsula
pentágono, pentagon
pentano, pentane
penumbra, penumbra
pepsina, pepsin
péptico, peptic, peptide
peptidasa, peptidase
peptización, peptization
percebe, barnacle
percentil, percentile
percibir, perceive

perclorato, perchlorate
pérdida de calor, heat loss
perenne, perennial
perennifolio, evergreen
perfecto, perfect
perfil, profile
perfil del suelo, soil profile
perforación, drilling
perfume, perfume
pericardio, pericardium
peridotita, peridotite
perigeo, perigee
perihelio, perihelion
perilla, knob
perímetro, perimeter
periódico[1], journal
periódico[2], periodic
período, period
Período Cuaternario, Quaternary Period
período de identidad, identity period
período de incubación, incubation period
período de gestación, gestation period
período de latencia, latent period
Período Devónico, Devonian Period
Período Ordovícico, Ordovician Period
Período Pérmico, Permian Period
período refractario, refractory period
periscopio, periscope
peristalsis, peristalsis
peritoneo, peritoneum
perjudicial, detrimental
perla[1], pearl
perla[2], bead
permafrost, permafrost
permanece sin cambios, remain the same
permanecer, remain
permanente, stable
permanentemente, permanently
permanganato de potasio, potassium permanganate
permeabilidad, permeability
permeabilidad selectiva, selective permeability
permeable, permeable
Pérmico, Permian
permitir, enable, permit, allow
perno, bolt
peroné, fibula
peróxido, peroxide
peróxido de hidrógeno, hydrogen peroxide
perpendicular, perpendicular
perpendicularmente, perpendicularly
perro, dog
perspectiva, perspective
perturbación, disturbance
perturbar, disturb
pesado, heavy
pescar[1], hook
pescar[2], fish
peso, weight
peso atómico, atomic weight
peso avoirdupois, avoirdupois weight
peso molecular, molecular weight
pestañas, eyelashes
peste, pest

pesticida, pesticide
pétalo, petal
petrificación, petrifaction
petróleo, petroleum
petrología, petrology
petroquímica, petrochemical
pez, fish
  peces planos, flatfish
  pez pulmonado, lungfish
pezón, nipple
pH, pH
  pH-metro, pH meter
pi, pi
picea, spruce
pico[1], beak, bill (bird)
pico[2], pico
picofaradios, picofarad
pie, foot (pl. feet)
  pie de atleta, athlete's foot
piedra, stone
  piedra angular, cornerstone
  piedra caliza, limestone
  piedra pómez, pumice
piel[1], skin
piel[2], fur
pierna, leg
piezoelectricidad, piezoelectricity
pigmento, pigment
  pigmento biliar, bile pigment
  pigmento complementario, complementary pigment
  pigmento primario, primary pigment
  pigmento secundario, secondary pigment
pila[1], battery
  pila galvánica, galvanic cell
  pila líquida, wet cell
  pila seca, dry cell
pila[2], pile, stack
  pila atómica, atomic pile
  pila de compost, compost pile
  pila de composta, compost pile
píldora, pill
piña, pine cone
pineal, pineal
pingüino, penguin
pinnado, pinnate
pino, pine
pinocitosis, pinocytosis
piñón, pine nut
pinta, pint
pinzas para tubo de ensayo, crucible tongs
pinzones, finches
piojo, louse
pionero, pioneer
pipeta[1], pipette
  pipeta de transferencia, transfer pipette
  pipeta graduada, graduated pipette
  pipeta volumétrica, volumetric pipette
pipeta[2], medicine dropper
pirámide, pyramid (math)
  pirámide alimenticia, food pyramid
  pirámide de energía,

energy pyramid
pirámide de la energía, pyramid of energy
pirámide ecológica, ecological pyramid
piridoxina, pyridoxine
pirita, pyrite
pirólisis, pyrolysis
piroxeno, pyroxene
pirrol, pyrrole
Piscis, Pisces
pista, raceway, racetrack, track
pistilo, pistil
pistola, pistol, gun
pistola láser, radar gun
pistón, piston
pitagórico, Pythagorean
pitón, python
pitónido, python
píxel, pixel
pizarra, slate
placa, plaque, board
placa base, motherboard (computer)
placa de circuito, circuit board
placa de la célula, cell plate
placa de Petri, Petri dish
placa neural, neural plate
placa continental, continental plate
placa ecuatorial, equatorial plate
placas de la Tierra, Earth's plates
placas litosféricas, lithospheric plates
placebo, placebo
placel, placer
placenta[1], placenta
placenta[2], afterbirth
plaga[1], plague
plaga[2], blight
plagioclasa, plagioclase
plan de acción para emergencias, emergency action plan
plan de generación de energía, energy generation plan
plan de investigación, research plan
plancha de vapor, steam iron
plancton, plankton
planeador, glider
planeta, planet
planeta enano, dwarf planet
planetas gaseosos gigantes, gas giants
planetas terrestres, terrestrial planets
plano[1], plane
plano de simetría, plane mirror
plano de polarización, plane polarized
plano inclinado, inclined plane
plano[2], flat
planta, plant
planta bienal, biennial plant
planta carnosa, succulent plant
planta cultivada, cultivated plant
planta insectívora, insectivorous plant

planta leguminosa, leguminous plant
planta suculenta, succulent plant
planta vascular, vascular plant, tracheophyte
planta verde, green plant
plantas no vasculares, nonvascular plant
plantilla, template
plaqueta, platelet
plaqueta en sangre, blood platelet
plasma, plasma
plasma sanguíneo, blood plasma
plásmido, plasmid
plasmodium, plasmodium
plástico, plastic
plastidio, plastid
plastificante, plasticizer
plasto, plastid
plata, silver
plataforma[1], shelf
plataforma continental, continental shelf
plataforma[2], platform
platelmintos[1], platyhelminthes
platelmintos[2], flatworm
platillo, cymbal
platino, platinum
plato, plate
plato de porcelana sin esmaltar, unglazed porcelain plate
plato teórico, theoretical plate
playa, beach
pleamar, flood tide
plegado, folded
Pleistoceno, Pleistocene
pleura, pleura
pleuronectiformes, flatfish
plexo, plexus
plexo solar, solar plexus
pliegue, fold
Plioceno, Pliocene
plomo, lead
pluma, plume, quill
plumaje, plumage
plumón, down (feather)
plúmula, plumule
Plutón[1], Pluto
plutón[2], pluton
plutonio, plutonium
pluvial, pluvial
población, population
poder de resolución de una lente, resolution power of lens
polar, polar
polaridad, polarity
polaridad invertida, reversed polarity
Polaris, Polaris
polarización, polarization
polarizado, polarized
polarizador, polarizer
polarizar, polarize
polarográfica, polarographic
polea, pulley
polea fija, fixed pulley
polea móvil, movable pulley
polen, pollen
poli-, poly-
polialcohol, polyhydric alcohol
policarbonato, polycarbonate
policíclicos, polycyclic

policloropreno, polychloroprene
policondensación, polycondensation
policromático, polychromatic
poliedro, polyhedron
polielectrolito, polyelectrolyte
poliéster, polyester
poliestireno, polystyrene
polietilenglicoles, polyethylene glycols
polietileno, polyethylene
poligénica, polygenic
poliglicol, polyglycol
polígono, polygon
poliinsaturado, polyunsaturated
poliisopreno, polyisoprene
polilla, moth
polimerización, polymerization
  polimerización térmica, thermal polymerization
polímero, polymer
  polímero alto, high polymer
  polímeros de condensación, condensation polymer
polimineralica, polymineralllic
polimorfismo, polymorphism
polinización, pollination
  polinización anemófila, wind pollination
  polinización cruzada, cross-pollination
  polinización por el viento, wind pollination
polinizar, pollinate
polinomio, polynomial
poliomielitis, polio
polipéptido, polypetide
poliploidía, polyploidy
pólipo, polyp
polipropileno, polypropylene
polisacárido, polysaccharide
politetrafluoroetileno, polytetraflouro ethylene
poliuretano, polyurethane
polivinílico, polyvinyl chloride
polluelo, fledgling
polo, pole
  polo celeste, celestial pole
  polo magnético, magnetic pole
  Polo Norte, North Pole
  Polo Sur, South Pole
  polos geográficos, geographic poles
polonio, polonium
polución, pollution
polvillo radiactivo, fallout
polvo[1], dust
  polvo radiactivo, atomic dust
polvo[2], powder
  polvo para hornear, baking powder
poner en circulación, circulate
por ciento, percent
por lo tanto, hence
porcelana, porcelain
porcentaje, percentage, percent
  porcentaje de error, percent error

porcentaje de la composición, percentage composition
porcentaje de la masa, percentage by mass
porción, portion
pórfido, porphyry
poríferos, porifera
poro, pore
porométrico, porometric
porosidad, porosity
poroso, porous
portador, carrier
posar, settle
poseer, possess
posición[1], position
posición aparente de las constelaciones, apparent positions of the constellations
posición de equilibrio, equilibrium position
posición[2], standing
positivo, positive
positrón, positron
posterior, posterior
postulado, postulate
postura erecta, erect posture
potasa, potash
potasio, potassium
potencia, power
potencia en vatios, wattage
potencial, potential
potencial de oxidrilo, pOH
potencial de polarización, polarization potential
potencial de reposo, resting potential
potencial eléctrico, electric potential
potencial hídrico, water potential
potencial normal de electrodo, standard electrode potential
potencial normal de oxidación-reducción, standard oxidation-reduction potential
potencial de ionización, ionization potential
potenciómetro, potentiometer
poza de marea, tidal pool
pozo artesiano, artesian well
pozo de mina, pit
prácticamente, practically
práctico, practical
pradera, grassland, prairie
prado, meadow
praliné, praline
praseodimio, praseodymium
precesión, precession
presión arterial, blood pressure
precipitación, precipitation
precipitado, precipitate
precipitante, precipitant
precisamente, precisely
precisión, precision, accuracy
preciso, precise
predador, predator
predecir, predict
predicción, prediction
predominar, predominate
prefijo, prefix
prefijo de centésima, centi
prefijo de decámetro, deka, deca
pregunta, question
pregunta de opción múltiple, multiple

choice question
prelavado, presoak
preliminar, preliminary
premolar, bicuspid, premolar
prensil, prehensile
preparación para emergencias, emergency preparedness
preparar, prepare
prepolímero, prepolymer
presa, prey
presencia, presence
preservar, preserve
preservativo, preservative
presión, pressure, strain, stress
presión alta, high pressure
presión atmosférica, atmospheric pressure, air pressure
presión atmosférica normal, standard atmospheric pressure
presión sanguínea, blood pressure
presión baja, low pressure (air front)
presión barométrica, barometric pressure
presión crítica, critical pressure
presión de aire, air pressure
presión de raíz, root pressure
presión de saturación, vapor pressure
presión de succión, suction pressure
presión de turgencia, turgor pressure
presión de vapor, vapor pressure
presión de vapor a saturación, saturation vapor pressure
presión diastólica, diastolic pressure
presión estándar, standard pressure
presión osmótica, osmotic pressure
presión parcial, partial pressure
presión sistólica, systolic pressure
presumiblemente, presumably
presunción, assumption
prevención, prevention
prevenir, prevent
previamente, previously
previo, previous
primario, primary
primate, primate
primavera, spring
primer meridiano, prime meridian
primer trimestre, first quarter
primera generación filial, first filial generation
Primera Ley de Movimiento de Newton, Newton's First Law of Motion
primera ley del movimiento, first law of motion
Primera Ley de la Termodinámica, First Law of Thermodynamics
primitivo, primitive
principal, major
principalmente, primarily
principio, principle

principio de Arquímedes, Archimedes principle
principio de De Broglie, De Broglie Principle
principio de exclusión competitiva, competitive exclusion principle
principio de incertidumbre, uncertainty principle
principio de incertidumbre de Heisenberg, Heisenberg uncertainty principle
principio de la horizontalidad original, principle of original horizontality
principio de Pascal, Pascal Law
principio de superposición, principle of superposition
principio del uniformismo, principle of uniformitarianism
prisma, prism
probabilidad, probability
probabilidad de ocurrencia, probability of occurrence
probable, probable
probar, prove
probeta, graduated cylinder
problema de aprendizaje, learning disability
problemas de carácter interdisciplinario, interdisciplinary problems
probóscide, proboscis
procariota, prokaryotic
procedimiento, procedure
procesamiento electrónico de datos, electronic data processing
proceso, process
proceso dinámico, dynamic process
procesos bioquímicos, biochemical processes
producir, generate
productividad, productivity
productividad primaria, primary productivity
producto, product
producto iónico, ion-product
producto parcial, partial product
productos competitivos, competing products
productor, producer
profase, prophase
progesterona, progesterone
programa, program
programable, programmable
progresión aritmética, arithmetic progression
prohibir, prohibit
promedio, average, mean
prometeo, promethium
prominencia, prominence
prominente, prominent
promontorio, cape
promotor, promoter
pronóstico, forecast
pronóstico del tiempo, weather forecasting
propagar, spread
propano, propane

propiedad, property
propiedad asociativa, associative property
propiedad física, physical property
propiedad química, chemical property
propiedades coligativas, colligative properties
propileno, propylene
propilo, propyl
proponer, propose
proporción, proportion
proporcional, proportional
proporcionalidad, proportionality
proporcionalidad constante, proportionality constant
propuesta, proposal
propulsión a chorro, jet propulsion
prosencéfalo, forebrain
prostaglandina, prostaglandin
próstata, prostate
protactinio, protactinium
proteasa, protease
proteger, protect
proteína, protein
proteína incompleta, incomplete protein
proteína completa, complete protein
protesta, protest
protista, protist
protocolo, protocol
protón, proton
protoplasma, protoplasm
protozoario, protozoan
protozoo (pl. protozoos), protozoan (pl. protozoa)
protrombina, prothrombin
proveer, provide
proximidad, proximity
proyección cónica, conic projection
proyección homolosena, homolosine projection
proyección sinusoidal, sinusoidal projection
proyectar, screen
proyectil, projectile
proyecto, project
prueba[1], trial
prueba[2], proof
prueba[3], test
prueba de biuret, biuret test
prueba xantoproteica, xanthoproteic test
psicología, psychology
psicosis, psychosis
psicrómetro, psychrometer
psila, sucker (zoology)
psiquiatría, psychiatry
pterodactiloideo, pterodactyl
ptialina, ptyalin
pubertad, puberty
pubis, pubis
puente salino, salt bridge
puerto (computadora), port (computer)
pulga, flea
pulga de agua, water flea
pulgada, inch
pulgar, thumb
pulmón, lung
pulmonar, pulmonary
pulpa, pulp
pulpo, octopus
pulsación, beat
púlsar, pulsar
pulso, pulse

pulso incidente, incident pulse
puma, mountain lion
punto, point
punto cardinal, cardinal point
punto ciego, blind spot
punto crítico, tipping point
punto de condensación al vacío, vacuum condensing point
punto de congelamiento, freezing point
punto de ebullición, boiling point
punto de ebullición normal, normal boiling point
punto de fusión, melting point
punto de inflamación, flash point
punto de referencia, benchmark, reference point
punto de rocío, dew point
punto de saturación, saturation point
punto decimal, decimal point
punto focal, focal point
punto focal principal, principal focal point
punto límite, tipping point
punto triple, triple point
puntos de coincidencia, common ground
puntos en común, commonalities, common ground
pupa, chrysalis
pupila, pupil (eye)
pureza, purity
purificación del agua, water purification
purina, purine
puro, pure
puro dominante, pure dominant
puro recesivo, pure recessive
púrpura, purple
pus, pus

quark, quark
quásar, quasar
quebrar, break
quelato, chelate
quemador de laboratorio, laboratory burner
quemadura, burning
quemar, burn
queratina, keratin
queroseno, kerosene
quijada, jaw bone
química, chemistry
química analítica, analytical chemistry
química forense, forensic chemistry
química inorgánica, inorganic chemistry
química orgánica, organic chemistry
químico, chemical

quimioautótrofos, chemoautotroph
quimiosíntesis, chemosynthesis
quimioterapia, chemotherapy
quimiotrofo, chemotroph
quimisorción, chemisorption
quimo, chyme
quinina, quinine
quino, cinchona
quinona, quinone
quiropráctica, chiropractic
quiste, cyst
quitar, remove
quitina, chitin

R

rabia, rabies
rabo (fruit), stalk (n)
racimo, raceme
radar, radar
radiación, radiation
  radiación adaptativa, adaptive radiation
  radiación electromagnética, electromagnetic radiation
  radiación cósmica, cosmic radiation
  radiación terrestre, terrestrial radiation
radiactividad, radioactivity
  radiactividad artificial, artificial radioactivity
radiactivo, radioactive
radiador, radiator
radial, radial
radialmente, radially
radián, radian
radiante, radiant
radical, radical
  radical libre, free radical
radicando, radicand
radícula, radicle
radio[1], radio
  radiodifusión, broadcast
radio[2] (elemento químico), radium
  radio atómico, atomic radius
radio[3] (geometría), radius
radio[4] (hueso), radius
radiocarbono, radiocarbon
radiofrecuencia, radio frequency
radioisótopo, radioisotope
radiología, radiology
radiómetro, radiometer
radiotelescopios, radio telescope
radioterapia, radiotherapy
radón, radon
ráfaga, squall
raíz, root
  raíz capilar, root hair
  raíz cuadrada, square root
  raíz cúbica, cube root
  raíz del diente, tooth root
  raíz principal, tap root
  raíz primaria, primary root
  raíz ventral, ventral root
rama (árbol), branch (tree)

ramita, twig
rampa, ramp
rana, frog
rango[1], range
rango[2], rank
ranura, slot (computer)
rápidamente, rapidly
rápido, quick
raqueta, racquet
raquitismo, rickets
rarefacción, rarefaction
raro, rare
rascar, scrape
rasgar, rip, tear
rasgo, feature, trait
  rasgo dominante, dominant trait
  rasgo en vías de extinción, disappearing trait
  rasgo hereditario, inherited trait
raspadura, scrape
raspar, scrape
rastro, track
rasuradora, razor
ratón[1], mouse (computer)
ratón[2], mouse (pl. mice)
rayar, scratch
rayo (física), ray
  rayo beta, beta ray
  rayo gama, gamma ray
  rayo vascular, vascular ray
  rayos catódicos, cathode ray
  rayos directos, direct rays
  rayos verticales, vertical rays
  rayos X, x-ray
raza[1] (animal), breed
raza[2], pedigree
razón[1], reason
razón[2], rate
  a razón de, rate
razonamiento científico, scientific thinking
reabsorción, reabsorption
reacción, reaction
  reacción de combustión directa, direct combustion reaction
  reacción de condensación, condensation reaction
  reacción en cadena, chain reaction
  reacción en cadena de la polimerasa, polymerase chain reaction (PCR)
  reacción nuclear, nuclear reaction
  reacción química, chemical reaction
  reacción de protólisis, protolysis reaction
  reacción de sustitución, substitution reaction
  reacción dependiente de la luz, light-dependent reaction
  reacción heterogénea, heterogeneous reaction
  reacción homogénea, homogeneous reaction
  reacción redox, redox reaction
  reacción reversible,

reversible reaction
reacción termonuclear, thermonuclear reaction
reacciones alérgicas, allergic reactions
reacciones de oscuridad, dark reaction
reaccionar, react
reactancia inductiva, inductive reactance
reactivo[1], reactant
reactivo[2], reagent
reactivo analítico, analytical reagent
reactor, reactor
reactor de fusión, fusion reactor
reactor nuclear, nuclear reactor
reactor reproductor, breeder reactor
reactores de fisión, fission reactor
reafirmación, restatement
real, actual, real
realista, realistic
realizar[1], perform
realizar[2], accomplish
reanimación cardiopulmonar (RCP), CPR
reborde, ledge
rebotar, bounce
rebote[1], bounce
rebote[2], rebound
recarga, recharge
receptáculo, receptacle
receptor[1], receiver
receptor universal, universal recipient
receptor[2], receptor
receptor sensorial, sensory receptor
recesivo, recessive
recibir, receive
reciclar, recycle
reciclaje, recycling
recientemente, recently
recipiente, container
recíproco, reciprocal
reclamar[1], claim
reclamar[2], reclaim
recolección directa, direct harvesting
recolección y procesamiento de información, gathering and processing information
recolectar, collecting
recombinación, recombination
recombinación cromosómica, chromosomal recombination
recombinación de gametos, recombination gamete
recombinar, recombine
reconocer, recognize
reconocimiento médico, physical examination
recopilar, collect, collecting
recopilar información, collect information
recorrido anual de las constelaciones, annual traverse of the constellations
recorte, clipping
recristalización, recrystallization
rectangular, rectangular
rectángulo, rectangle
rectificación, rectification
rectilíneo, rectilinear

recto[1], rectum
recto[2], straight
recuento de células sanguíneas, blood count
recuperar, recover
recurso, resource
  recurso energético no renovable, nonrenewable energy resource
  recurso energético renovable, renewable energy
  recurso natural, natural resource
  recurso no renovable, nonrenewable resource
  recurso renovable, renewable resource
red, grid
  red alimenticia, food web
  red cristalina, crystal lattice
  red de comunicaciones electrónicas, electronic communications network
  red nerviosa, nerve net
  red potencial, net potential
redefinir, redefine
reducción, reduction
  reducción de la llama, reducing flame
reducido, reduced
reducir, reduce
reemplazar, replace
reemplazo, replacement
referencia, reference
  referencias bibliotecarias, library references
  referencias electrónicas, electronic references
referente, concerning
referir, refer
refinación, refining
refinería, refinery
reflejar, reflect
reflejo, reflex
  reflejo condicionado, conditioned reflex
  reflejo rotuliano, knee-jerk reflex
reflexión, reflection
  reflexión difusa, diffuse reflection
  reflexión regular, regular reflection
  reflexión total interna, total internal reflection
reflexionar, reflect
reflujo[1], ebb (tide)
reflujo[2], reflux
reforestación, reforestation
reforzar, reinforce
refracción, refraction
  refracción de ondas, wave refraction
refractar, refract
refrigerador, refrigerator
refrigerante, refrigerant
refuerzo, reinforcement
refutar[1], refute
refutar[2], disprove
regeneración, regeneration
región, region
  region de elongación, elongation region
  región geografía, source region
registrar, record

registro, register
regla de cálculo, slide rule
regla de la mano izquierda, left-hand rule
regla graduada, ruler
regla métrica, metric ruler
regolito, regolith
regulación, regulation
regulado, reduction division (meiosis)
regulador, dimmer
regular[1], regular
regular[2], regulate
regularizado, regulated
regurgitación, regurgitation
reina, queen
reino, kingdom
  reino de las plantas, Plantae
  reino Protista, Protista
  reino vegetal, vegetable kingdom
rejilla, grating
  rejilla de difracción, diffraction grating
relación[1], ratio
relación[2], relationship
  relación de edad, age relationship
  relación depredador-presa, predator-prey relationship
  relación parasitaria, parasitic relationship
  relaciones intersectoriales, cross-cutting relationships
relacionado, related
relacionar, relate
relámpago, lightning
relativamente, relatively
relatividad, relativity
relé, relay
relevante, relevant
relleno, filler
  rellenos sanitarios, sanitary landfills
reloj atómico, atomic clock
reloj biológico, biological clock
reloj de sol, sundial
remache, rivet
remisión, remission
remolino[1], eddy
remolino[2], swirl
remolino[3], whirlpool
rémora, remora
remoto, remote
renacuajo, tadpole
renal, renal
renina, rennin
renio, rhenium
renovable, renewable
reordenamiento, rearrangement
reordenar, rearrange
reorganización, rearrangement
reóstato, rheostat
repelente, repellent
repentinamente, suddenly
repetir, repeat
repetitiva, repetitious
réplica, aftershock
replicar, replicate
repollo, cabbage
represa, dam
representar, represent
reproducción, reproduction, breeding
  reproducción asexual, asexual reproduction
  reproducción de esporas, spore reproduction

reproducción selectiva, selective breeding
reproducción sexual, sexual reproduction
reproducción vegetativa, asexual reproduction
reproducir, reproduce
reptil, reptile
repulsión, repulsion
repulsivo, repulsive
requerido, required
resaca[1], backwash
resaca[2], undertow
resaltar, accentuate
reserva genética, gene pool
resfriado común, common cold
residuo[1] (matemáticas), remainder
residuo[2], residue
residuos líquidos, runoff
residuo[3], waste
residuos nitrogenados, nitrogenous waste
residuos peligrosos, hazardous waste
residuos sólidos, solid waste
resina, resin
resina epoxica, epoxy resin
resina sintética, synthetic resin
resistencia, resistance
resistencia de la roca, rock resistance
resistencia del aire, air resistance
resistencia efectiva, effective resistance
resistente a, resistant to
resistir, resist
resistor, resistor
resolución vectorial, vector resolution
resolver, resolve
resonancia, resonance
resonancia magnética nuclear (RMN), magnetic resonance imaging (MRI)
resorte, spring
respaldo, support
respectivo, respective
respiración[1], breathing
respiración[2], respiration
respiración celular, cellular respiration
respiración celular, cell respiration
respiración externa, external respiration
responder, respond
responsable, responsible
respuesta, response
respuesta inflamatoria, inflammatory response
respuesta inmune, immune response
resta, subtraction
restablecer, restore, reestablish
restaurar, restore
restos, remains
resultado, result
resultante, resultant
resultar, result
resumir, summarize
retícula, reticulum
reticular, reticulate
retina, retina
retirar, scratch
retoño, shoot

retorta, retort
retransmitir, relay
retroalimentación, feedback
    retroalimentación negativa, negative feedback
    retroalimentación positiva, positive feedback
retrocruzamiento, test cross
retrógrado, retrograde
retrovirus, retrovirus
reunidos, assembled
reunir, collect
reutilizar, reuse
revelar, reveal
reverso, reverse
revestimiento del útero, uterine lining
revisados por colegas, peer reviewed
revisados por pares, peer reviewed
revisión, review
    revisión literaria, literature review
    revisión por pares, peer review
revolución, revolution
    revolución verde, green revolution
riboflavina, riboflavin
ribosa, ribose
ribosoma, ribosome
riel de metal, metal runner
riesgo, risk
rigidez, stiffness
rígido, rigid
rigor mortis, rigor mortis
rinoceronte, rhinoceros
riñón, kidney
río, river
riolita, rhyolite
ritmo, beat
    ritmo biológico, biorhythm
    ritmo circadiano, circadian rhythm
rizoide, rhizoid
rizoma, rhizome
RMN (resonancia magnética nuclear), MRI (magnetic resonance imaging)
roble, oak
robot, robot
robótica, robotics
roca, rock
    roca clástica, clastic rock
    roca ígnea, igneous rock
    roca ígnea intrusiva, intrusive igneous rock
    roca metamórfica, metamorphic rock
    roca no sedimentaria, nonsedimentary rock
    roca parental, parental rock
    roca primaria, primary rock
    roca rodado, boulder
    roca sedimentaria, sedimentary rock
rociar, spray
rocío, dew
rod (unidad de longitud), rod
rodante, rolling
rodear, surround
rodilla, knee
rodio, rhodium
roedor, rodent

rojo, red
Rojo Congo, Congo Red
rojo neutro, neutral red
rol, role
rollo, roll
ROM (memoria de solo lectura), ROM (read-only memory)
rombencéfalo, hindbrain
rombo[1], rhombus
rombo[2], diamond
romper[1], snap
romper[2], break
rompevientos, wind break
rotación, rotation
rotación de cultivos, crop rotation
rotación de la Tierra, Earth's rotation
rotación óptica, optical rotation
rotación de cultivos, rotation of crops
rotar, rotate
rotífero, rotifer
rótula, patella
router, router
rubéola[1], German measles
rubéola[2], rubella
rubí, ruby
rubidio, rubidium
rudimentario, vestigial
rueda y eje, wheel and axle
ruidoso, loudness
rumiante, ruminant
ruptura, hernia
rutenio, ruthenium
rutina, routine

sabana, savanna
sabio, wise
sacarasa, sucrase
sacárido, saccharide
sacarina, saccharin
sacarosa, sucrose, saccharose
saco[1], sac
saco embrionario, embryo sac
saco vitelino, yolk sac
saco[2], sack
sacro, sacrum
sacudida, shock
Sagitario, Sagittarius
sal, salt
sal de Epsom, Epsom salts
sal de mesa, table salt
sal de piedra, rock salt
sal de soda, sal soda
sal fundida, fused salt
sal normal, normal salt
salamandra, salamander
salamandra acuática, newt
salida[1], output
salida[2], outlet
salina[1], salt mine
salina[2], salt marsh
salinidad, salinity
salino, saline
saliva, saliva

salivar, salivate
salmón, salmon
salmonella, salmonella
salmuera, brine
saltación, saltation
saltamontes, grasshopper
salud, health
samario, samarium
sangre, blood
  (de) sangre caliente, warm blooded
  sangre entera, whole blood
  sangre Rh negativo, Rh negative blood
  sangre Rh positivo, Rh positive blood
sanguijuela, leech
sanitario, sanitary
sapo, toad
saponificación, saponification
saprofitismo, saprophytism
saprófito, saprophyte
sarampión, measles
sarcoma, sarcoma
sarpullido, heat rash
satélite, satellite
  satélite artificial, artificial satellite
saturación, saturation
saturado, saturated
Saturno, Saturn
savia, sap
saxófono, saxophone
schirrus, cirro
sebo, sebum
secante, secant
secar, desiccate
sección transversal[1], cross-section
sección transversal[2], transverse section
seco, dry
secreción, secretion
secretar, secrete
secretina, secretin
sector, sector
secuencia, sequence
  secuencia principal, main sequence
secundario, secondary
seda, silk
sedante, sedative
sedativo, sedative
sedimentación, sedimentation
  sedimentación ordenada, graded bedding
  sedimentación potencial, sedimentation potential
sedimentario, sedimentary
sedimento, sediment
  sedimento residual, residual sediment
  sedimento transportado, transported sediment
segmentación (biología), cleavage
segmento, segment
segregación, segregation
según, according to
segundo(a), second
  Segunda Ley de Movimiento de Newton, Newton's Second Law of Motion
  Segunda Ley de la Termodinámica, Second Law of Thermodynamics
  segunda regla de la mano izquierda, second left-hand rule

segunda generación filial, second filial generation
selección, selection
selección natural, natural selection
selección artificial, artificial selection
seleccionar, select
selecto, select
selenio, selenium
sellador, sealant
selva, jungle
selva tropical, rain forest
semana, week
semejanza, similarity
semen, semen
semiárido, semiarid
semicapa, subshell
semicírculo, semicircle
semiconductor, semiconductor
semilla, seed
semimicroanálisis, semimicroanalysis
semipermeable, semipermeable
semirreacción, half-reaction
semirrecta (matemáticas), ray
semisintético, semisynthetic
seno, breast, sine, sinus
sensación, sensation
sensibilidad, sensitivity
sensibilización, sensitization
sensible, sensitive
sensor, sensor
sensorial, sensory
sentido[1], sense
sentido[2], direction
sentido antihorario, counterclockwise
sentido horario, clockwise
sentir, sense
sépalo, sepal
separación[1], separation
separación[2], ablation
separado, separated
separar, separate
sepsis, sepsis
sequía, drought
ser no vivo, nonliving thing
serie, series
serie de los lantánidos, lanthanide series
serie de parafina, paraffin series
serie de transición, transition series
serie electromotriz, electromotive series
serie del benceno, benzene series
serie radiactiva, radioactive series
series homólogas, homologous series
serpentear, meander
serpiente, snake
servidor, server
servir como, served as
sésil, sessile
seta, mushroom
seudópodo, pseudo pod
sexo, sex
sexual, sexual
SIDA (síndrome de inmunodeficiencia adquirida), AIDS (acquired immunodeficiency syndrome)
sidéreo, sidereal
siembra de nubes, cloud seeding

siempreverde, evergreen
sierra, mountain chain, sierra
sífilis, syphilis
sifón, siphon
significativo, significant
sílex, chert, flint
silicato, silicate
sílice, silica
silicón, silicon
silicona, silicone
Siluriano, Silurian
siluro, catfish
simbiosis, symbiosis
simbiótico, symbiotic
símbolo, symbol
simetría, symmetry
  simetría bilateral, bilateral symmetry
  simetría radial, radial symmetry
simétrico, symmetrical
simio, ape
simpático, sympathetic
simple, simple
simplificar, simplify
sin cambio, unchanged
sin clasificar, unsorted
sin disponibilidad, unavailability
sin fricción, frictionless
sinapsis, synapse
sinclinal, syncline
sincrociclotrón, synchrocyclotron
sincrotón, synchrotron
sindicato, union
síndrome, syndrome
  síndrome de Down, Down syndrome
  síndrome de inmunodeficiencia adquirida (SIDA), acquired immunodeficiency syndrome (AIDS)
  síndrome de alcoholismo fetal, fetal alcohol syndrome
  síndrome del túnel carpiano, carpal tunnel syndrome
  síndrome Turner, Turner's syndrome
  síndrome de Klinefelter, Klinefelter's syndrome
sinnúmero, countless
sinónimo, synonym
sinterizar, sinter
síntesis, synthesis
  síntesis por deshidratación, dehydration synthesis
sintético, synthetic
sintetizar, synthesize
síntoma, symptom
sirena, siren
Sirio, Sirius
sísmico, seismic
sismógrafo, seismograph
sismología, seismology
sistema, system
  sistema aislado, isolated system
  sistema cardiovascular, cardiovascular system
  sistema cerrado de energía, closed energy system
  sistema ciclónico, cyclonic system
  sistema circulatorio abierto, open circulatory system
  sistema circulatorio cerrado, closed circulatory system

sistema de alerta temprana, early warning system
sistema de coordenadas, coordinate system
sistema de coordinación, coordinating system
sistema de erosión y depósito, depositional—erosional system
sistema de impresión, offset
sistema de numeración binaria, binary number system
sistema de órganos, organ system
Sistema de Posicionamiento Global (GPS), Global Positioning System (GPS)
sistema de raíces fibrosas, fibrous root system
sistema de transporte, transporting system
sistema decimal, decimal system
sistema endocrino, endocrine system
sistema gastrointestinal, gastrointestinal tract
sistema geológico dinámico, dynamic geologic system
sistema hidráulico, hydraulic system
sistema inmunitario, immune system
sistema inmunológico, immune system
Sistema IUPAC, IUPAC system
sistema linfático, lymphatic system
sistema métrico, metric system
sistema muscular, muscular system
sistema nervioso, nervous system
sistema nervioso central, central nervous system
sistema nervioso parasimpático (SNP), parasympathetic nervous system (PNS)
sistema nervioso periférico, peripheral nervous system
sistema nervioso simpático, sympathetic nervous system
sistema nervioso somático, somatic nervous system
sistema operativo, operating system
sistema químico, chemical system
sistema reproductor femenino, female reproductive system
sistema solar, solar system
sistema tetragonal, tetragonal system
sistema urinario, urinary system
sistema vascular,

vascular system
sistemas terrestres, Earth Systems
sistemático, systematic
sístole, systole
sitio, site
sitio activo, active site
sitio web, website (computer)
situación, situation
situación de erosión, erosional situation
sobras, remains
sobreabundancia, overabundance
sobrecarga, overload
sobrecruzamiento, crossing-over
sobredosis, drug overdose
sobrepastoreo, overgraze
sobrepesca, overfishing
sobrepoblación, overpopulation
sobres unidosis de café, coffee pods
sobresaturación, supersaturation
sobrevivir, survive
sobrevivir, surviving
sociología, sociology
soda, soda
sodio, sodium
software, software
software de dominio público, public domain software
sol[1], sun
sol[2], sol
solar, solar
soldadura, brazing
soleado, sunny
solenoide, solenoid
solera, ground-sill
solidificación, solidification
solidificar, solidify
sólido, solid
Sólidus, solidus
soliquoide, soliquoid
solsticio, solstice
solsticio de invierno, winter solstice
solsticios, solstices
soltar, release
solitaria, tapeworm
solubilidad, solubility
soluble, soluble
solución, solution
solución concentrada, concentrated solution
Solución de Bennedict, Benedict's solution
solución de equilibrio, solution equilibrium
solución del suelo, soil solution
solución diluida, diluted solution
solución estándar, standard solution
solución neutra, neutral solution
solución normal, normal solution
solución saturada, saturated solution
solución sobresaturada, supersaturated solution
solución sólida, solid solution
solución verdadera, true solution
soluto, solute
solvente, solvent
solvente latente, latent solvent

sombra de lluvia, rain shadow
someterse a, undergo
sonambulismo, sleepwalking
sonámbulo, sleepwalker
sónar, sonar
sonda, probe
sónico, sonic
sonido, sound
soporte del tubo de prueba, test tube holder
soprano, soprano
sorción, sorption
sospechar, suspect
sostener, maintain
sotavento, leeward
spirogyra, spirogyra
spoiler, spoiler
stratus, stratus
suave, smooth
subatómico, subatomic
subclase, subclass
subconjunto, subset
subcutáneo, subcutaneous
subespecie, subspecies
subfilo, subphylum
subir, rise
  subir el cursor, scroll up
sublimación, sublimation
sublimar, sublimate
sublime, sublime
submarino, submarine
subnivel, sublevel
subproducto, byproduct
subreino, subkingdom
substracción, subtraction
subsuelo, subsoil
subtropical, subtropical
succión, suction
sucesión, succession, sequence
  sucesión de Fibonacci, Fibonacci sequence
  sucesión ecológica, ecological succession
  sucesión primaria, primary succession
  sucesión secundaria, secondary succession
suceso impactante, impact event
sudor, sweat
sueco, Swedish
suelo, soil, topsoil
  suelo ácido, acid soil
  suelo alcalino, alkaline soil
  suelo arenoso, sandy soil
  suelo de morrena, ground moraine
  suelo maduro, mature soil
  suelo residual, residual soil
  suelo transportado, transported soil
sueño, sleep
suero, serum
  suero sanguíneo, blood serum
suficientemente, sufficiently
sulfato, sulfate
  sulfato de bario, barium sulfate
  sulfato de calcio, calcium sulfate
  sulfato de cromo, chromium sulfate
  sulfato ferroso, ferrous sulfate
sulfuro, sulfide
  sulfuro de cadmio, cadmium sulfide

sulfuro de hidrógeno, hydrogen sulfide
suma, addition, sum
suma de cocientes, ratio sum
suma vectorial, vector sum
sumando, addend
sumergir[1], engulf
sumergir[2], immerse
sumergir[3], submerge
sumergirse[1], dip
sumergirse[2], plunge
sumersión, submergence
sumidero (energía), sink (energy)
sumidero de baja temperatura, low temperature sink
superar, overcome
superávit, surplus
superabundancia, overabundance
superconductor, superconductor
supercontinente, super continent
superenfriamiento, supercooling
superficial, shallow
superficie de las vías respiratorias, respiratory surface
superficie de reacción, surface reaction
superficie de sustentación, airfoil
superficie química, surface chemistry
supergigante, supergiant
supergravedad, super gravity
supernova, supernova
superóxidos, superoxides
supersimetría, super symmetry
supersónico, supersonic
sobrevivir, survive
supervivencia, survival
supervivencia del más apto, survival of the fittest
suplementación, supplementation
suponer, assume
suponiendo, assuming
suposición, assumption
supresión, deletion
surcos, grooves
surcos en el lecho de roca, bedrock grooves
surgir, emerge
suspender, suspend
suspensión, suspension
sustancia, substance
sustancia blanca, white matter
sustituto, substitute
sustractivo, subtractive
sustraendo, subtrahend
sustrato, substrate
sutura, suture
suturar, stitch
Syndet, syndet

tabaco, tobacco
tabique nasal, septum
tabla de datos, data table
tabla periódica, periodic table
Tablas de Referencia para las Ciencias de la Tierra, Earth Science Reference Tables

TAC (tomografía axial computarizada), CAT scan (Computed Axial Tomography)
táctil, tactile
tacto, touch
taiga, taiga
tálamo, thalamus
talasemia, thalassemia
talco, talc
talio, thallium, stem
tallo (plant), stalk (n)
 tallo leñoso, woody stem
talón, heel
talud continental, continental slope
talus, talus
tamaño, size
también, likewise
tambor, drum
tamiz molecular, molecular sieve
tangente, tangent
tanque, tank
tántalo, tantalum
tapíz, carpet
tapón de dos orificios, two-hole stopper
tapón de un orificio, one-hole stopper
tarántula, tarantula
tarifa angular, angular rate
tarso, tarsal
tasa de aceleración, rate of acceleration
tasa metabólica basal, basal metabolic rate
tasa húmeda adiabática, moist adiabatic lapse rate
Tauro, Taurus
taxonomía, taxonomy
TDA (trastorno por déficit de atención), ADD (Attention Deficient Disorder)
tecla, key (computer)
 tecla control, control key (computer)
 tecla de bloqueo de mayúsculas, capital lock key (computer)
 tecla de mayúsculas, shift key (computer)
 tecla de servicio, control key (computer)
 tecla enter, enter key (computer)
 tecla inicio, home key (computer)
 tecla suprimir, delete key (computer)
teclado, keyboard (computer)
tecnecio, technetium
técnica, technique
tecnología, technology
 tecnología de la información, information technology
tecnólogo, technologist
tectita, tektite
tectónica de placas, plate tectonics
tejido, tissue
 tejido adiposo, adipose tissue
 tejido conectivo, connective tissue
 tejido conjuntivo, connective tissue
 tejido de almacenamiento, storage tissue
 tejido esponjoso, spongy tissue

tejido foliar, lamina
tejido muscular, muscle tissue
tejido óseo, bone tissue
tejido sanguíneo, blood tissue
tejido vascular, vascular tissue
tejo, yew
tela, web (spider)
telaraña, spiderweb
telecomunicación, telecommunication
telégrafo, telegraph
telemetría, telemetry
telencéfalo, cerebrum
telescopio, telescope
telescopio reflector, reflection telescope
telescopio refractor, refracting telescope
telofase, telophase
telurio, tellurium
telururo, telluride
temblar, shiver
temblor, tremor
temperatura, temperature
temperatura absoluta, absolute temperature
temperatura crítica, critical temperature
temperatura del aire, air temperature
templado, temperate
temporalmente, temporary
tenazas[1], pincers
tenazas[2], tongs
tendencia[1], tendency
tendencia[2], trend
tender, tend
tender a, tend to
tendón, tendon
tendón de Aquiles, Achilles tendon
tendón de la corva, hamstring
tenedor, fork
tenia, tapeworm
tensión, tension, strain
tensión superficial, surface tension
tensoactivo, surfactant
tentáculo, tentacle
teorema, theorem
teorema de Pitágoras, Pythagorean theorem
teoría, theory
teoría calórica, caloric theory
teoría celular, cell theory
teoría cuántica, quantum theory
teoría de cuerdas, string theory
teoría de las placas, plate theory
teoría de los gérmenes de la enfermedad, germ theory of disease
teoría de la generación espontánea, spontaneous generation theory
teoría del campo unificado, unified field theory
teoría general de la relatividad, general theory of relativity
teoría científica, scientific theory

teoría cinética molecular, kinetic molecular theory
teoría de la relatividad, theory of relativity
teoría del uso y desuso, theory of use and disuse
teorías actuales, current theories
terabyte, terabyte
terapia, therapy
terapia de choque, shock therapy
terapia de electrochoque, shock therapy
terapia génica, gene therapy
terbio, terbium
Tercera Ley de Movimiento de Newton, Newton's Third Law of Motion
terciario, tertiary
termes, termite
térmico, thermal, heat
terminal, terminal
término, term
termión, thermion
termita, thermite
termo, vacuum bottle
termobalanza, thermobalance
termocupla, thermocouple
termodinámica, thermodynamics
termoeléctrico, thermoelectric
termoendurecibles, thermosetting
termoformado, thermoform
termógrafo, thermograph
termómetro, thermometer
termómetro de bulbo seco, dry bulb thermometer
termonuclear, thermonuclear
termoplástico, thermoplastic
termoquímica, thermochemistry
termósfera, thermosphere
termostato, thermostat
termoterapia, heat treatment
ternario, ternary
ternero, calf
terraza, terrace
terraza marina, marine terrace
terremoto, earthquake
terreno inundable, floodplain
terrestre, terrestrial
terrígeno, terrigenous
territorio, territory
Terylene, Terylene
Tesla, Tesla
testículo, testicle, testis
testosterona, testosterone
teta, teat
tétanos, tetanus
tetera, kettle
tetracloruro, tetrachloride
tetracloruro de carbono, carbon tetrachloride
tétrada, tetrad
tetraedro, tetrahedron
tetraedro de silicio-oxígeno, silicon-oxygen tetrahedron
tetrafluoroetileno, tetraflouroethylene
tetrágono, tetragonal
tetrámero, tetramer
tetraploide, tetraploid

textura, texture
textura de suelo, soil texture
tiamina, thiamine
tibia, tibia
tibio, warm
tiburón, shark
tiempo, time
tiempo medio, mean time
tiempo medio de Greenwich, Greenwich Mean Time
tiempo solar, solar time
Tierra[1], Earth
tierra[2], earth, ground
tierra[3], land
tierra firme, dry land
tierras baldías, badlands
tierras raras, rare earth
tierras de pantanos, wetlands
tierras, terrae
tifoidea, typhoid
tifón, typhoon
tifus, typhus
tijeras, scissors
tijereta, earwig
timbre, timbre
timina, thymine
timo, thymus, thymus gland
tímpano, eardrum
tinción, staining
tinta, ink
tintura, tincture
tiña, ringworm
tiol, thiol
tipo[1], type
tipo[2], sort
tipo de parásito, fluke
tipo de suelo, soil texture
tipo de suelo rojo, Indian Red
tiranosaurio, tyrannosaurus
tiras bimetálicas, bimetallic strip
tiritar, shiver
tiro muy fuerte, cannonball
tiroides, thyroid
tirosina, tyrosine
tiroxina, thyroxine
titanio, titanium
titulación por precipitación, precipitation titration
tizón, smut
tlacuache, opossum
tobillo, ankle
tocar, touch
todavía, still
todo, entire
tolerar, tolerate
tolueno, toluene
toma de decisiones del consumidor, consumer decision making
toma de huellas dactilares, fingerprinting
tómbolo[1], baymouth bar
tómbolo[2], tombolo
tomografía axial computarizada (TAC), Computed Axial Tomography (CAT scan)
tonelada, ton
tóner, toner
tono[1], pitch
tono[2], hue
tonsila, tonsils
topacio, topaz
topo, mole

topografía, topography
topografía kársticas, karst topography
topología, topology
toque, touch
tórax, thorax
torcedura[1], strain
torcedura[2], twist
torio, thorium
tormenta de nieve, blizzard
tormenta de polvo, dust storm
tormenta eléctrica, thunderstorm
tormenta magnética, magnetic storm
tornado, tornado
tornasol, litmus
tornillo, screw
toro (matemáticas), torus
torre, tower
torsión, torsion
tortuga, tortoise, turtle
tos ferina, whooping cough
tosco, rough
totora, cattail
tóxico, toxic
toxicología, toxicology
toxina, toxin
trabajador portuario, dockhand
trabajar, work
trabajo, work
tracción, tensile
tracto, tract
tractor, tractor
tráfico, traffic
tragar, swallow
trago[1], drink
trago[2], swallow, gulp
trampolín, trampoline
tranquilizante, tranquilizer
transacciones conjuntas, joint ventures
transcripción, transcription
transducción, transduction
transductor, transducer
transferencia, transfer
transferencia de calor, heat transfer, heat exchange
transferir, transfer
transformación, transformation
transformación de límites, transform boundaries
transformador, transformer
transformador reductor, step-down transformer
transformador elevador, step-up transformer
transformar, transform
transfusión, transfusion
transfusión sanguínea, blood transfusion
transgénico, transgenic
transición, transition
transición demográfica, demographic transition
transistor, transistor
translocación, translocation
translúcido, translucent
transmisión, transmission
transmisible (enfermedad), communicable
transmisor, transmitter
transmitir, transmit
transmutación, transmutation
transparencia atmosférica, atmospheric transparency
transparente, transparent
transpiración[1], perspira-

tion, sweat
transpiración[2], transpiration
transpirar, sweat
transplante, transplant
transplante de corazón, heart transplant
transportador, protractor
transportar, transport
transporte, transport
transporte activo, active transport
transversales, cross-cutting
transverso, transverse
trap en electrónica, trap number
trapecio, trapezium
trapezoide, trapezoid
tráquea, trachea, windpipe
trastorno por déficit de atención (TDA), Attention Deficient Disorder (ADD)
tratamiento, treatment
tratamiento de aguas residuales, sewage treatment
tratar, treat
trauma, trauma
trayectoria, trajectory
trayectoria aparente, apparent path
trazador, tracer
trematodo, fluke
trementina, turpentine
tren, train
traqueófita, tracheophyte, vascular plant
tríada, triad
triangulación, triangulation
triángulo, triangle
triángulo isósceles, isosceles
triángulo rectángulo, right triangle
tributario, tributary
tríceps, triceps
triceratops, triceratops
trichloroflouromethane, trichloroflouromethane
tricíclicos, tricyclic
triclínico, triclinic
tricloroetileno, trichloroethylene
tricúspide, tricuspid
trietanolamina, triethanolamine
triethylaluminum, triethylaluminum
triglicérido, triglyceride
trigo, wheat
trigonometría, trigonometry
trillón, trillion
trilobite, trilobite
trinchera, trench
trineo, sled
trinitrotolueno, TNT
trióxido, trioxide
trióxido de arsénico, arsenic trioxide
triploide, triploid
tripsina, trypsin
triquinelosis, trichinosis
triquinosis, trichinosis
trismo, lockjaw
tritio, tritium
tritón, newt
trivalente, trivalent
tromba marina, waterspout
trombina, thrombin
trombo, thrombus
trombón, trombone
tromboplastina, thromboplastin
trombosis, thrombosis
trompa de Eustaquio, Eustachian tube

trompa de Falopio, Fallopian tube
trompeta, trumpet
tronco[1], trunk (tree)
 tronco cerebral, brain stem
tronco[2], log
tropical, tropical
trópico, tropic
 trópico de Cáncer, Tropic of Cancer
 trópico de Capricornio, Tropic of Capricorn
tropismo, tropism
tropósfera, troposphere
trucha arco iris, rainbow trout
trueno, thunder
trypanosomatida, trypanosome
tsunami, tsunami
tubérculo, tuber
tuberculosis, tuberculosis
tubería[1], pipe
tubería[2], tubing
tubo, tube
 tubo bronquial, bronchial tube
 tubo de cardo, thistle tube
 tubo de ensayo, test tube
 tubo de entrega, delivery tube
 tubo de polen, pollen tube
 tubo de rayos catódicos, cathode-ray tube
 tubo de vacío, vacuum tube
 tubo de fermentación, fermentation tube
 tubo digestivo, digestive tract, alimentary canal
 tubo Geiger Müller, Geiger-Müller tube
túbulo, tubule
 túbulos seminíferos, seminiferous tubule
tuétano, marrow
tufo, tuff
tulio, thulium
tumor, tumor
tundra, tundra
túnel, tunnel
 túnel de viento, wind tunnel
tungsteno[1], tungsten
tungsteno[2], wolfram
tunicado, tunicate
turba, peat
turbidez, turbidity
turbiditas, turbidites
turbina, turbine
 turbinas de viento, wind turbines
turborreactor, turbojet
turbulento, turbulent
turgente, turgid

U

ubicación, location
ubre, udder
úlcera, ulcer
 úlcera péptica, peptic ulcer
última instancia, ultimate analysis
ultracentrífuga, ultracentrifuge
ultrafiltración, ultrafiltration

ultrasonido, ultrasound
ultravioleta, ultraviolet
umbela, umbel
umbra, umbra
umbral, threshold
ungulado, ungulate
unicelular, unicellular
único, unique
unidad astronómica, astronomical unit
unidad central de procesamiento (CPU), central processing unit (CPU)
unidad de disco, disk drive
unidad de disco duro, hard drive
unidad de fuerza, unit
unidad de masa atómica, atomic mass unit
unidad derivada, derived unit
unidad internacional, international unit
unidades de fuerza, units of force
uniforme, uniform
uniformemente, uniformly
unión neuromuscular, neuromuscular junction
unión[1], union
unión[2], joint
univalente, univalent
univalvo, univalve
universo, universe
  universo abierto, open universe
uña, fingernail, nail
  uña del pie, toenail
uracilo, uracil
uranio, uranium
Urano, Uranus
urbanización, urbanization
urea, urea
ureasa, urease
uretra[1], ureter
uretra[2], urethra
urgencia, rush
urticaria, hives
uso, usage, use
  uso de la tierra, land use
  uso indebido de drogas, drug abuse
  uso sustentable, sustainable use
utensilio, utensil
útero[1], uterus
útero[2], womb
utilizar, utilize
úvula, uvula

vacante[1], vacancy
vacante[2], vacant
vaciar, hollow
vacío[1], empty
vacío[2], vacuum
vacuna, vaccine
  vacuna de poliovirus inactivados, Salk vaccine
  vacuna Sabin, Sabin vaccine
  vacuna Salk, Salk vaccine
vacunación, vaccination
vacunado, vaccinated
vacuola, vacuole
  vacuolas alimenticias, food vacuole
vagina, vagina
vaina[1], pod

vaina[2], hull, husk
 vaina de mielina, myelin sheath
 vaina de la hoja, leaf sheath
valencia, valence
válida, valid
valina, valine
valiosa, valuable
valle, valley
 valle colgante, hanging valley
 valle del Rift, Rift Valley
 valle en forma de U, U-shaped valley
 valle en forma de V, V-shaped valley
 valle glaciar, valley glacier
valor, value
 valor absoluto, absolute value
 valor adaptativo, adaptive value
 valor observado, observed value
 valor X, X-value
 valor Y, Y-value
valoración, titration
 valoración potenciométrica, potentiometric titration
valorante, titrant
válvula, valve
 válvula bicúspide, bicuspid valve
 válvula de corredera, slide value
 válvula electrónica, electronic tube
 válvula semilunar, semilunar valve
vanadio, vanadium
vapor[1], vapor
 vapor de agua, water vapor
 vapor de presión negativa, vapor pressure depression
vapor[2], steam
vaporización, vaporization
vaporizar, vaporize
variabilidad, variability
variable, variable
 variable controlada, controlled variable
 variable dependiente, dependent variable
 variable independiente, independent variable
 variable manipulada, manipulated variable
 variables atmosféricas, atmospheric variable
 variables originarias, original variables
variación, variation
 variación genética, genetic variation
 variación inversa, inverse variation
 variaciones climáticas, climate variations
varicela, chickenpox
variedad[1], variety
variedad[2] (planta), breed
varilla métrica, meter stick
vascular, vascular
vaso[1], glass
 vaso de agua, water glass
vaso[2], vessel

vaso sanguíneo, blood vessel
vaso sanguíneo ventral, ventral blood vessel
vasos linfáticos, lymph vessel
vaso de precipitados, beaker
vaso de precipitado graduado, graduated beaker
vasoconstricción, vasoconstriction
vasodilatación, vasodilation
vasopresina, vasopressin
vástago, scion, shoot (n)
vataje, wattage
vatio, watt
vatio-hora, watt-hour
vecinal, vicinal
vector[1], vector
vector[2], vectorial
vestigios, vestiges
vegetaciones, adenoids
vegetal, vegetable
vehículo, vehicle
vejez, old age
vejiga, bladder, urinary bladder
vejiga natatoria, swim-bladder
vela, candle
veleta, weather vane, wind vane, vane
vellosidad, villi
vellosidades, villus
velocidad[1], rate
velocidad[2], speed, velocity
velocidad angular, angular velocity
velocidad de escape, escape velocity
velocidad de la luz, velocity of light
velocidad de la onda, wave velocity
velocidad de las olas, wave speed
velocidad de reacción, reaction rate
velocidad final, final velocity
velocidad inicial, initial velocity
velocidad instantánea, instantaneous speed, instantaneous velocity
velocidad media, average speed
velocidad orbital, orbital speed, orbital velocity
velocidad promedio, average velocity
velocidad terminal, terminal velocity
vena, vein, vena
vena cava, vena cava
vena cava superior, superior vena cava
vena porta renal, renal portal vein
vena pulmonar, pulmonary vein
vena renal, renal vein
venación, venation
vencer, overcome
veneno[1], venom
veneno[2], poison
venenoso, poisonous

venera, scallop
ventaja, advantage
ventaja mecánica, mechanical advantage
ventaja mecánica ideal (IMA), ideal mechanical advantage (IMA)
ventisca de nieve, blizzard
ventral, ventral
ventrículo, ventricle
vénula, venule
Venus, Venus
Venus atrapamoscas, Venus flytrap
ver, see
verano, summer
verdadero, real
verde, green
verificación, verification
verruga, wart
vértebra, vertebra
vértebras caudales, caudal vertebrae
vertebrado, vertebrate
vertedero de basura, landfill
verter, dump
vertical[1], upright
vertical[2], standing
verticalmente, vertically
vértice, vertex (pl. vertices), apex
vertidos nucleares, nuclear waste
vesícula, vesicle
vesícula biliar, gallbladder
vesícula pinocítica, pinocytic vesicle
vesícula seminal, seminal vesicle
vestigial, vestigial
veta[1], streak
veta[2], seam
(en) vez de, instead
vía, track, path
vía de combustión, combustion path
Vía Láctea, Milky Way
vías urinarias, urinary tract
vías respiratorias, respiratory tract
vibración, vibration
vibracional, vibrational
vibrar, vibrate
vida, life
vida media, half-life
vidrio, glass
viento[1], wind
vientos alisios, trade winds
viento fuerte, high winds
viento solar, solar wind
vientos prevalecientes, prevailing winds
vientos planetarios, planetary winds
viento[2], windward
vigor heterosis, hybrid vigor
vigor híbrido, hybrid vigor
vigorosamente, vigorously
VIH (virus de la inmunodeficiencia humana), HIV (human immunodeficiency virus)
vinagre, vinegar
vinculación, bonding
vinculación electrovalente, electrovalent bonding
vinilo, vinyl

vino tinto, red wine
violar, violate
Virgo, Virgo
virtual, virtual
(en) virtud de, by virtue of
viruela, smallpox
virus, virus
  virus de Ébola, Ebola virus
  virus de la inmunodeficiencia humana (VIH), human immunodeficiency virus (HIV)
  virus libre, free-living virus
víscera, viscera
viscosidad, viscosity
viscoso, viscous
visible, visible
visualizar, visualize
vital, vital
vitamina, vitamin
vítreo, vitreous
viuda negra, black widow
vivíparo, viviparous
vivisección, vivisection
vivo, living
vocal[1], vowel
vocal[2], vocal
volátil, volatile
volatilizar, volatilize
volcán, volcano
  volcán compuesto, composite volcano
  volcán de cono de ceniza, cinder cone volcano
  volcán en escudo, shield volcano
voltaico, voltaic
voltaje, voltage
  voltaje de polarización, bias voltage
voltímetro, voltmeter
voltio, volt
volumen[1], volume
  volumen de una botella, volume bottle
  volumen gramo-molecular, gram-molecular volume
volumen[2], loudness
volumétrico, volumetric
voluntario[1], volunteer
voluntario[2], voluntary
vórtice, vortex, whirlpool
vulcanización, vulcanization
vulcanizar, vulcanize
vulva, vulva

W-Z

Web[1], Web (computer)
web[2], tela (spider)
wolframio, wolfram
World Wide Web, World Wide Web
xantófila, xanthophyll
xenón, xenon
xilema, xylem
xileno, xylene
yacimiento (geología), deposit
yarda, yard (measurement)
yegua, mare
yema[1], bud
  yema terminal, terminal bud
yema[2], yolk
yeso, gypsum
yeyuno, jejunum
yodo, iodine
yoduro, iodide

yoduro de hidrógeno, hydrogen iodide
yuyo, weed
zafiro, sapphire
zángano, drone
zanja, trench
zarcillo, tendril
zarigüeya, opossum
zarpa, claw
zeolita, zeolite
zodiaco, zodiac
zona, zone
zona afótica, aphotic zone
zona caliente, hot spot
zona de aireación, zone of aeration
zona de baja presión, trough, low pressure area
zona de calmas ecuatoriales, doldrums
zona de convergencia, convergence zone, zone of convergence
zona de divergencia, zone of divergence, divergence zone
zona de elongación, elongation zone
zona de fractura, rift zone
zona de refinación, refining zone
zona de saturación, zone of saturation
zona de subducción, subduction zone
zona de transición, transition zone
zona fótica, photic zone
zona horaria, time zone
zona polar, polar zone
zona rift, rift zone
zona templada, temperate zone
zona tropical, tropical zone
zonificación, zonation
zoología, zoology
zoológico, zoo
zooplancton, zooplankton
zorro, fox
zumbido, ringing

Section II
English - Spanish

abacus, ábaco
abandon, abandonar
abbreviation, abreviatura
abdomen, abdomen
aberration, aberración
ability, capacidad, aptitud
abiogenesis, abiogénesis
abiotic factor, factor abiótico
ablation, extirpación, separación, ablación
abomasum, abomaso
abortion, aborto
abrade, corroer, erosionar
abrasion, abrasión
abscess, absceso
abscissa, abscisa
absence, ausencia
absolute, absoluto
  absolute altitude, altitud absoluta
  absolute humidity, humedad absoluta
  absolute magnitude, magnitud absoluta
  absolute scale, escala absoluta
  absolute temperature, temperatura absoluta
  absolute value, valor absoluto
  absolute zero, cero absoluto
absorb, absorber
absorption, absorción
  absorption spectrum, espectro de absorción
abyssal plains, llanura abisal
accelerate, acelerar
acceleration, aceleración
  acceleration of gravity, aceleración de la gravedad
accentuate, acentuar, resaltar
access, acceso
accompany, acompañar
accomplish, realizar, llevar a cabo
accordance, conforme a
according to, según
account, explicación
accumulate, acumular
accuracy, precisión, exactitud
acetaldehyde, etanal, acetaldehído
acetate, acetato
acetic acid, ácido acético
acetone, acetona
acetylcholine, acetilcolina
acetylene, acetileno
acetylsalicylic acid, ácido acetilsalicílico
achieve, lograr, conseguir
Achilles tendon, tendón de Aquiles
achromaticity, acromaticidad
acid, ácido
  acid rain, lluvia ácida
  acid soil, suelo ácido
acidity, acidez
acne, acné
acoustics, acústica
acquire, adquirir
acquired characteristics, características adquiridas
acquired immunity, inmu-

nidad adquirida
acquired immunodeficiency syndrome (AIDS), síndrome de inmunodeficiencia adquirida (SIDA)
acromegaly, acromegalia
acrylic acid, ácido acrílico
actinide, actínido
actinium, actinio
action, acción
activation energy, energía de activación
active immunity, inmunidad activa
active site, sitio activo
active transport, transporte activo
activity, actividad
actual, real
actual evapotranspiration, evapotranspiración real
actual observations, observaciones reales
acupuncture, acupuntura
acute angle, ángulo agudo
adapt, adaptar
adaptation, adaptación
adaptive, adaptativo
adaptive radiation, radiación adaptativa
adaptive value, valor adaptativo
ADD (Attention Deficient Disorder), trastorno por déficit de atención (TDA)
addend, sumando
addiction, adicción
addition, suma
additional, adicional
additive, aditivo
adenine, adenina
adenoids, adenoides, vegetaciones
adenosine monophosphate (AMP), adenosín monofosfato
adenosine triphosphate (ATP), adenosín trifosfato
adequate, adecuado
adhesive, adhesivo
adiabatic, adiabático
adiabatic temperature change, cambio de temperatura adiabático
adipose, adiposo
adipose tissue, tejido adiposo
adjacent, adyacente
adjacent angle, ángulo adyacente
adjust, ajustar
adolescence, adolescencia
adrenal cortex, corteza suprarrenal
adrenal gland, glándula suprarrenal
adrenal medulla, médula suprarrenal
adrenaline, adrenalina
adrenocorticotropic hormone (ACTH), hormona adrenocorticotrópica
adsorption, adsorción
adsorption isobar, isóbaras de adsorción
adsorption isotherm, isotermas de adsorción
adult, adulto
advance, avanzar
advantage, ventaja

aerate, oxigenar, airear
aeration, aireación
aerobe, aerobio
aerobic, aerobio
aerodynamic, aerodinámico
aeronautics, aeronáutica
aerosol, aerosol
aerospace, aeroespacio
affect, afectar
afterbirth, placenta
aftershock, réplica
agar, agar, gelosa
agarose, agarosa
agate, ágata
age relationship, relación de edad
agent, agente
agglutinogen, aglutinógeno
aging, envejecimiento
agricultural, agrícola
AIDS (acquired immunodeficiency syndrome), SIDA (síndrome de inmunodeficiencia adquirida)
ailment, enfermedad
air, aire
- **air mass,** masa de aire
- **air pocket,** bolsa de aire
- **air pressure,** presión de aire, presión atmosférica
- **air resistance,** resistencia del aire
- **air sac,** alvéolo pulmonar
- **air space,** espacio aéreo
- **air temperature,** temperatura del aire
- **air track,** carril de aire

airborne molecule, molécula transportada por el aire
aircraft, avión
airfoil, superficie de sustentación
airplane, avión
airwaves, ondas de radio
albino, albino
albumen, albumen
albumin, albúmina
alchemy, alquimia
alcohol, alcohol
aldehyde, aldehídos
alga, alga
algal bloom, floración de algas
algebra, álgebra
algebraic, algebraico
algorithm, algoritmo
alicyclic hydrocarbons, hidrocarburos alicíclicos
alien (ecology), especie exótica
alimentary canal, tubo digestivo
aliphatic hydrocarbons, hidrocarburos alifáticos
alkalescence, alcalescencia
alkali, álcali
- **alkali metal,** metal alcalino

alkaline, alcalino
- **alkaline earth metal,** metal de tierra alcalina
- **alkaline soil,** suelo alcalino

alkaloid, alcaloide
alkane, alcano
- **alkane derivative,** alcano-derivados

alkene, alqueno

alkyl, alquilo
alkyne, alquino
allantosis, alantoides
allene, aleno
allergen, alérgeno
allergic, alérgico
  allergic reactions, reacciones alérgicas
allergy, alergia
alligator, caimán, aligátor
allotrope, alótropo
allotropy, alotropía
allow, permitir, admitir, asignar, aplicar
alloy, aleación
  alloy steel, aleación de acero
alluvial fan, abanico aluvial
alluvium, aluvión
alpha emission, emisión alfa
alpha particle, partícula alfa
alpine glacier, glaciar alpino
alter, modificar, alterar, cambiar
altered, modificado, alterado
  altered gene, gen alterado
alteration, alteración
alternate, alternar
  alternate angles, ángulos alternos
alternating current (A.C), corriente alterna (C.A.)
  alternating current generator, generador de corriente alterna
alternative medicine, medicina alternativa
alternator, alternador
altimeter, altímetro
altitude, altitud
  altitude of Polaris, altura de la Estrella Polar
alto, alto
alum, alumbre
alumina, alúmina
aluminum, aluminio
alveolus, alvéolo
Alzheimer's disease, enfermedad de Alzheimer, mal de Alzheimer
amalgam, amalgama
amber, ámbar
ambient, ambiente
ameba, amiba, ameba
ameboid movement, movimiento ameboide
amethyst, amatista
amino acid, aminoácido
amino group, grupo amino
ammeter, amperímetro
ammonia, amoníaco
  ammonia liquor, amoníaco líquido
ammonification, amonificación
ammonium, amonio
amnesia, amnesia
amniocentesis, amniocentesis
amnion, amnios
amoeba, amiba, ameba
amount, cantidad
ampere, amperio
amperometry, amperímetro
amphetamine, anfetamina
amphibian, anfibio
amphibole, anfíbol
amphiprotic, anfitrópico
amphoteric, anfótero
amphoterism, anfoterismo
amplification, amplificación
amplitude, amplitud
amyl alcohol, alcohol amílico
amylase, amilasa

anabolism, anabolismo, biosíntesis
anaconda, anaconda
anadromous, anádromo
anaerobic, anaerobio
analgesic, analgésico
analog, analógico
analogous, análogo
analogy, analogía
analysis, análisis
analyst, analista
analytic geometry, geometría analítica
analytical balance, balanza analítica
analytical chemistry, química analítica
analytical reagent, reactivo analítico
analyze, analizar
anaphase, anafase
anatomy, anatomía
ancestor, ancestro
ancient, antiguo, anciano
andesite, andesita
androgen, andrógeno
anemia, anemia
anemometer, anemómetro
aneroid, aneroide
  aneroid barometer, barómetro aneroide
anesthesia, anestesia
aneurysm, aneurisma
angina pectoris, angina de pecho
angiosperm, angiosperma
angle, ángulo
  angle of incidence, ángulo de incidencia
  angle of insolation, ángulo de insolación
  angle of reflection, ángulo de reflexión
  angle of refraction, ángulo de refracción
angstrom, ángstrom
angular momentum, momento angular
angular rate, tarifa angular
angular unconformity, discordancia angular
angular velocity, velocidad angular
anhydride, anhídrido
anhydrous, anhidro
animal, animal
Animalia, mundo animal, reino animal
anion, anión
ankle, tobillo
annual, anual
  annual traverse of the constellations, recorrido anual de las constelaciones
annular, anular
anode, ánodo
anomaly, anomalía
anorexia, anorexia
ant, hormiga
antacid, antiácido
antagonist, antagonista
Antarctic Circle, círculo polar antártico
Antarctic Continent, continente antártico
antenna, antena
anterior, anterior
anther, antera
anthracite, antracita
anthrax, ántrax, carbunclo, carbunco
anthropology, antropología
antibiotic, antibiótico
antibody, anticuerpo
antichlor, anticloro

anticline, anticlinal
anticyclone, anticiclón
antidote, antídoto
antienzyme, antienzima
antifebrin, Antifebrin
antifoaming agent, agente antiespumante
antifreeze, anticongelante
antigen, antígeno
antihistamine, antihistamínico
antilogarithm, antilogaritmo
antimony, antimonio
antinodal line, línea antinodal
antinode, antinodo
antioxidant, antioxidante
antiparticle, antipartícula
anti-return, antirretorno
antiseptic, antiséptico
antitoxin, antitoxina
antler, cuerno
anus, ano
aorta, aorta
aortic arch, arco aórtico
apatite, apatito
ape, simio
apex, vértice
aphelion, afelio
aphotic zone, zona afótica
apogee, apogeo
apparent, aparente
  apparent daily motion, movimiento diario aparente
  apparent diameter, diámetro aparente
  apparent magnitude, magnitud aparente
  apparent motion, movimiento aparente
  apparent motion of the planets, movimiento aparente de los planetas
  apparent path, trayectoria aparente
  apparent planetary diameter, diámetro aparente planetario
  apparent positions of the constellations, posición aparente de las constelaciones
  apparent solar day, día solar aparente
appendicitis, apendicitis
appendix (pl. appendices), apéndice
appliance, aparato eléctrico
apply, aplicar
appropriate, apropiado
approval, aprobación
approximate, aproximado
approximately, aproximadamente
aqua, agua
aquaculture, acuicultura
aquarium, acuario
aquatic, acuático
aquifer, acuífero
Arabic numeral, número arábigo
arachnid, arácnido
arc, arcos
arch, arco
archaeology, arqueología
archaeon, arquea
archer, arquero
Archimedes principle, principio de Arquímedes
architect, arquitecto
Arctic Circle, círculo polar ártico
area, área

arête, arista
argon, argón
arid, árido
Aristotle, Aristóteles
arithmetic, aritmética
  arithmetic mean, media aritmética
  arithmetic progression, progresión aritmética
arm, brazo
armadillo, armadillo
armature, armadura
aromatic, aromático
around, alrededor
arrange, ordenar
arrangement, acuerdo, combinacón
array, conjunto, matriz, arreglo
arrow, flecha
arsenic, arsénico
  arsenic trioxide, trióxido de arsénico
arsenical, arsénico
arsenide, arseniuro
arteriole, arteriola
arteriosclerosis, arteriosclerosis
artery, arteria
artesian well, pozo artesiano
arthritis, artritis
arthropod, artrópodo
artificial, artificial
  artificial intelligence, inteligencia artificial
  artificial radioactivity, radiactividad artificial
  artificial satellite, satélite artificial
  artificial selection, selección artificial
artillery, artillería
artiodactyls, artiodáctilo
asbestos, asbesto, amianto
ascending colon, colon ascendente
ascending flow, flujo ascendente
ASCII, código ASCII
ascorbic acid, ácido ascórbico
aseptic, aséptico
asexual, asexual
  asexual reproduction, reproducción asexual, reproducción vegetativa
asexually, asexualmente
ash, ceniza
aspect, aspecto
aspen, álamo temblón
asphalt, asfalto
asphyxia, asfixia
aspirate, aspirar
aspirin, aspirina
assemble, armar
assembled, reunidos
assimilation, asimilación
assistant, asistente
associate, asociar
association, asociación
  association neuron, neurona de asociación
associative property, propiedad asociativa
assume, suponer
assuming, suponiendo
assumption, presunción, suposición
aster, aster
asteroid, asteroide
asthenosphere, astenósfera
asthma, asma
astigmatism, astigmatismo

astringent, astringente
astrometry, astrometría
astronomer, astrónomo
astronomical, astronómico
astronomical unit, unidad astronómica
astronomy, astronomía
astrophysics, astrofísica
asymmetry, asimetría
asymptote, asíntota
atavistic, atávico
ataxia, ataxia
atherosclerosis, arteriosclerosis
athlete, atleta
athlete's foot, pie de atleta
atmosphere, atmósfera, ambiente
atmospheric, atmosférico
atmospheric carbon dioxide, dióxido de carbono atmosférico
atmospheric circulation, circulación atmosférica
atmospheric cross-section, muestra atmosférica
atmospheric data, datos atmosféricos
atmospheric pressure, presión atmosférica
atmospheric transparency, transparencia atmosférica
atmospheric variable, variables atmosféricas
atoll, atolón
atom, átomo
atomic, atómico
atomic absorption spectrometry, espectrometría de absorción atómica
atomic bomb, bomba atómica
atomic clock, reloj atómico
atomic dust, polvo radiactivo
atomic energy, energía atómica
atomic mass, masa atómica
atomic mass unit, unidad de masa atómica
atomic number, número atómico
atomic pile, pila atómica
atomic radius, radio atómico
atomic weight, peso atómico
atria, aurículas
atrium, aurícula
atrophy, atrofia
attain, conseguir, alcanzar
Attention Deficient Disorder (ADD), TDA (trastorno por déficit de atención)
attitude, actitud
attract, atraer
attraction, atracción
attractive, atractivo
audience, audiencia
auditorium, auditorio
auditory, auditivo
auditory canal, conducto auditivo
auditory nerve, nervio auditivo, vestibulococlear

augite, augita
auricle, aurícula
aurora, aurora
  aurora borealis, aurora boreal
autism, autismo
autoclave, autoclave
autoimmune, autoinmune
automatic, automático
automobile, automóvil
autopsy, autopsia
autosome, autosoma
autotrophic, autótrofo
  autotrophic nutrition, nutrición autótrofa
autumn, otoño
autumnal equinox, equinoccio de otoño
auxin, auxina
availability, disponibilidad
available, disponible
avalanche, alud, avalancha
average, promedio
  average speed, velocidad media
  average velocity, velocidad promedio
avian, aviar
aviary, aviario, pajarera
avoirdupois weight, peso avoirdupois
axil, axila
axiom, axioma
axis, eje
axle, eje
axon, axón
azeotropic, azeotrópica
azimuth, azimut
azurite, azurita, malaquita azul

baboon, babuino, papión
back, espalda, lomo
backbone, columna vertebral, espina dorsal
backwash, reflujo, resaca
bacteria, bacterias
bacterial, bacteriana
  bacterial infection, infección bacteriana
  bacterial pneumonia, neumonía bacteriana
bacteriophage, bacteriófago
bacterium, bacteria
badlands, tierras baldías, abarrancamiento
baffle, deflector
baking powder, levadura en polvo, polvo para hornear
baking soda, bicarbonato de sodio
balance[1], balanza
balance[2], equilibrio
balance[3], balancear
balanced diet, dieta balanceada
balanced forces, fuerzas balanceadas
baleen, barba de ballena
ball and socket joint, articulación de rótula
ballistics, balística
bamboo, bambú
band, ligamento
banding, bandas elásticas
bar graph, gráfico de barras
barb, lengüeta
bare, desnudo, descubierto

barite, baritina, barita
barium, bario
barium carbonate, carbonato de bario
barium chloride, cloruro de bario
barium nitrate, nitrato de bario
barium sulfate, sulfato de bario
bark, corteza
barnacle, percebe
barometer, barómetro
barometric pressure, presión barométrica
barrel, barril
barrier, barrera
barrier beach, cordón litoral
barrier island, isla barrada
barrier reef, barrera de coral
barycenter, baricentro
baryon, bariónico
basal metabolic rate, tasa metabólica basal
basalt, basalto
base, base
base level, nivel inferior
base pair, par de bases
based on which, a partir de la cual
basic, básico
basic anhydride, anhídrido básico
basic equation, ecuación básica
basin, cuenca
basis, base, fundamento
bat, murciélago
batholith, batolito
bathymetric map, mapa batimétrico
bathyscaphe, batiscafo
batter, golpear
battery, pila, batería
bauxite, bauxita
bay, bahía
baymouth bar, tómbolo
bayou, bayou, pantano
beach, playa
beach face, cordón costero
bead, perla
beak, pico
beaker, vaso de precipitados
beam balance, báscula mecánica
bear, oso
beat, ritmo, latido, pulsación
beaver, castor
becquerel, becquerel
bed, cama
bedding, cama, lecho, estratificación
bedload, carga de fondo
bedrock, lecho rocoso
bedrock grooves, surcos en el lecho de roca
bee, abeja
beetle, escarabajo
behave, comportarse
behavior, conducta, comportamiento
behavioral, comportamiento
behind, detrás
bell curve, curva normal, campana de Gauss
benchmark, punto de referencia
bend, doblar, curvar
beneath, debajo

Benedict's solution, Solución de Bennedict
beneficial, beneficioso
benefit, beneficio
benign, benigno
bent, curvatura
benthos, bentos
benzaldehyde, benzaldehído
benzene, benceno
  benzene ring, anillo bencénico
  benzene series, serie del benceno
benzenoid, benzenoide
benzidine, bencidina
benzoic acid, ácido benzoico
benzyl alcohol, alcohol bencílico
beriberi, beriberi
berkelium, berkelio
berm, berma
berry, baya
beryl, berilo
beryllium, berilio
best, el mejor, la mejor, lo mejor
beta carotene, beta-caroteno
beta particle, partícula beta
beta ray, rayo beta
beyond, más allá
bias voltage, parcialidad de polarización, voltaje de polarización
bicarbonate indicator, indicador de bicarbonato
biceps, bíceps
bicuspid, premolar
  bicuspid valve, válvula bicúspide
biennial, bienal
biennial plant, planta bienal
Big Bang, (el) Big Bang, (la) gran explosión
Big Dipper, Osa Mayor
bilateral symmetry, simetría bilateral
bile, bilis
  bile duct, conducto biliar
  bile pigment, pigmento biliar
bill (bird), pico
billiard, billar
bimetallic, bimetálico
  bimetallic strip, tiras bimetálicas
binary, binario
  binary code, código binario
  binary digit, dígito binario
  binary fission, fisión binaria
  binary number system, sistema de numeración binaria
  binary star, estrella binaria, estrella doble
bind, atar, ligar
binder, carpeta
binocular, binocular
binomial, binomio
biochemical, bioquímico
  biochemical processes, procesos bioquímicos
biochemistry, bioquímica
biodegradable, biodegradable
biodiversity, biodiversidad
bioethics, bioética
biogenesis, biogénesis
biogeography, biogeografía
biological activity, actividad biológica
biological catalysts, catalizadores biológicos

biological clock, reloj biológico
biological control, control biológico
biological magnification, magnificación biológica
biology, biología
biomass, biomasa
biomechanics, biomecánica
bionics, biónica
biophysics, biofísica
biopsy, biopsia
bioremediation, biorremediación
biorhythm, biorritmo, ritmo biológico
biosphere, biósfera
biotechnological, biotecnológicas
biotechnology, biotecnología
biotic, biótico
  biotic factor, factor biótico
biotin, biotina
biotite, biotita
biped, bípedo
bipedal, bípedo
birch, abedul
bird, ave
birth canal, canal de parto
birth control, control de natalidad
bismuth, bismuto
bit, bit
bitmap, mapa de bits
bitumen, betún
bituminous coal, hulla
biuret test, prueba de biuret
bivalent, bivalente
bivalve, bivalvo
black, negro
  black hole, agujero negro
  black dwarf, enana negra
  black widow, viuda negra
bladder, vejiga
blade, brizna, limbo
blastula, blástula
bleach, blanqueador
blending inheritance, herencia mezclada
blight, plaga
blind spot, punto ciego
blink[1]**,** parpadear
blink[2]**,** parpadeo
blizzard, tormenta, ventisca de nieve
block[1]**,** bloque
block[2]**,** bloquear
blog, blog, bitácora
blood, sangre
  blood cavity, cavidad sanguínea
  blood cell, glóbulo
  blood circulation, circulación sanguínea
  blood clot, coágulo
  blood count, recuento de células sanguíneas
  blood gas, gas sanguíneo
  blood group, grupo sanguíneo
  blood plasma, plasma sanguíneo
  blood platelet, plaqueta en sangre
  blood pressure, tensión, presión sanguínea, presión arterial

blood serum, suero sanguíneo
blood smear, frotis de sangre
blood sugar, azúcar en la sangre, glicemia
blood tissue, tejido sanguíneo
blood transfusion, transfusión sanguínea
blood type, grupo sanguíneo
blood vessel, vaso sanguíneo
blossom, florecer, flor
blowhole, orificio nasal
blowing agent, agente espumante
blubber, grasa de ballena
blue, azul
blueshift, corrimiento al azul
blur, difuminar
bob, inclinación
boil, hervir
boiler scale, escala de ebullición
boiling, ebullición, hirviente, hervor
boiling point, punto de ebullición
boiling point elevation, elevación del punto de ebullición
boldface, negrita
bolt, perno
Boltzmann constant, Constante de Boltzmann
bomb, bomba
bond, enlace
bond angle, ángulo de bonos
bonding, vinculación, aspecto afectivo
bone[1], hueso (animal)
bone tissue, tejido óseo
bone marrow, médula ósea
bone[2], espina (pez)
booklet, libreta, cuadernillo, folleto
Boolean algebra, álgebra booleana
borax, bórax
boreal, boreal
boric acid, ácido bórico
boroflouride, boroflouride
boron, boro
boron carbide, carburo de boro
boron hydride, hidruro de boro
boron nitride, nitruro de boro
borosilicate, borosilicato
botany, botánica
bottle, botella
boulder, roca, canto rodado
bounce[1], rebotar
bounce[2], rebote
boundary, límite
bovine, bovino
bow, arco
bowel, intestino
bowling ball, bola de boliche
Bowman's capsule, cápsula de Bowman
bowstring, cuerda de arco
box modeling, modelo de compartimentos
brachiopod, braquiópodo
brackish water, agua salobre
brain, cerebro

brain stem, tronco cerebral
brake fluid, líquido de frenos
branch (tree), rama (árbol)
branch-chain, cadena ramificada
brass, latón
brazing, soldadura
bread mold, pan de molde
break, romper, quebrar
breast, pecho, seno
breathing, respiración
breccia, brecha
breed, raza (animal), variedad (planta)
breeder reactor, reactor reproductor
breeding, reproducción, cría
brick, ladrillo, bloque
brief, breve
briefly, brevemente
bright line spectrum, espectro de línea brillante
brine, salmuera
bristle, cerda
brittle, frágil
broadcast, difusión
broccoli, brócoli
bromeliad, bromeliácea
bromide, bromuro
bromine, bromo
bromthymol blue, azul de bromotimol
bronchi, bronquios
bronchial tube, tubo bronquial
bronchiole, bronquiolo
bronchitis, bronquitis
bronchus, bronquio
bronze, bronce
brown dwarf, enana marrón, enana parda
browser, navegador
bubble, burbuja
bubble chamber, cámara de burbujas
bubo, bubón
bud, brote, yema, capullo
budding, incipiente
buffer, amortiguador
bug, bicho
bugle, clarín, corneta
bulb, bulbo
bulge, bulto
bulimia, bulimia
bullet, bala
bundle, paquete, manojo
Bunsen burner, mechero de Bunsen
buoyancy, flotabilidad
buoyant, flotante, boyante
buoyant force, fuerza boyante
burette, bureta
burglar, ladrón
burglar alarm, alarma antirrobo
burl, lupia
burn, quemar
burning, combustión, quemadura
butane, butano
butanediol, butanodiol
butanol, butanol
butene, buteno
butte, monte aislado
butterfly, mariposa
buttocks, nalgas
butyl, butílico
butyl alcohol, alcohol butílico
butyl rubber, caucho butílico
butyric acid, ácido butírico
by virtue of, en virtud de

bypass, bypass
byproduct, subproducto
byte, byte

cabbage, repollo
cable (computer), cable
cactus, cactus
cadmium, cadmio
  cadmium sulfide, sulfuro de cadmio
caffeine, cafeína
calcification, calcificación
calcine, calcinar
calcite, calcita
calcium, calcio
  calcium carbide, carburo de calcio
  calcium carbonate, carbonato de calcio
  calcium chloride, cloruro de calcio
  calcium hydroxide, hidróxido de calcio
  calcium hypochlorite, hipoclorito de calcio
  calcium oxide, óxido de calcio
  calcium phosphate, fosfato de calcio
  calcium sulfate, sulfato de calcio
calculate, calcular
calculator, calculadora
calculus, cálculo
calf, ternero, novillo
calibrate, calibrar
californium, californio
caliper, calibre
callus, callo
calomel, calomel
caloric, calórico
  caloric theory, teoría calórica
calorie, caloría
calorimeter, calorímetro
calorimetry, calorimetría
cambium, cámbium
Cambrian, Cámbrico
camera, cámara
camouflage, camuflaje
camp, acampar
camper, campista
camphor, alcanfor
cancer, cáncer
candela, candela
candle, vela
canine, canino
cannonball, tiro muy fuerte, bala de cañon
canoe, canoa
canopy, dosel
canyon, cañón
capable, capaz
capacitance, capacitación
capacitor, condensador
capacity, capacidad
cape, cabo, promontorio
capillarity, capilaridad
capillary, capilar
  capillary action, capilaridad, acción capilar
capital lock key (computer), tecla de bloqueo de mayúsculas
capsule, cápsula
capture, captar, captura
carapace, caparazón
carbene, carbeno
carbide, carburo
carbocyclic, carbocíclico
carbohydrase, carbohidrasa

carbohydrate, carbohidrato
carbolic acid, ácido carbólico
carbon, carbono
  carbon black, negro de carbón
  carbon cycle, ciclo del carbono
  carbon dating, datación por carbono
  carbon dioxide (CO_2), dióxido de carbono (CO_2)
  carbon disulfide, disulfuro de carbono
  carbon fixation, fijación del carbono
  carbon footprint, huella de carbono
  carbon grain, granos de carbono
  carbon monoxide, monóxido de carbono
  carbon tetrachloride, tetracloruro de carbono
carbonate, carbonato
carbonate carbonato, carbonato de carbono
carbonation, carbonatación
carbon-containing, contenedor de carbono
carbonic acid, ácido carbónico
carbonide, carbonido
carboniferous, carbonífero
carboxyl, carboxilo
  carboxyl group, grupo carboxilo
  carboxylic acid, ácido carboxílico
carburetion, carburación
carcinogen, carcinógeno, cancerígeno
cardiac, cardíaco
  cardiac muscle, músculo cardíaco
  cardinal number, número cardinal
  cardinal point, punto cardinal
cardiology, cardiología
cardiovascular disease, enfermedad cardiovascular
cardiovascular system, sistema cardiovascular
carnivore, carnívoro
carnivorous, carnívoro
carotene, caroteno
carotid artery, arteria carótida
carpal, carpo
  carpal tunnel syndrome, síndrome del túnel carpiano
carpel, carpelo
carpet, alfombra, moqueta, tapiz
carrier, portador
carrying power, llevar la energía
Carso, Carso
cartilage, cartílago
carton, caja de cartón
case, caso, asunto
cast, emitir, lanzar, arrojar
caste, casta
cat, gato
CAT scan (Computed Axial Tomography), TAC (tomografía axial computarizada)
catalase, catalasa
catalyst, catalizador
cataract, catarata
category, categoría
catenate, encadenar
caterpillar, oruga

catfish, siluro
catheter, catéter
cathode, cátodo
cathode ray, rayo catódico
cathode-ray tube, tubo de rayo catódico
cation, catión
cattail, enea, totora
caudal fin, aleta caudal
caudal vertebrae, vértebras caudales
caudatum, caudatum
cauliflower, coliflor
cause, causa
cave, cueva
cavitation, cavitación
cavity, cavidad
CD-ROM, CD-ROM
cecum, ciego (del intestino)
celestial, celeste
celestial equator, Ecuador celeste
celestial motions, movimientos celestes
celestial object, objeto celeste
celestial phenomenon, fenómeno celeste
celestial pole, polo celeste
celestial sphere, esfera celeste
cell[1], célula
cell plate, placa de la célula
cell-mediated immunity, inmunidad mediada por células
cell[2], celda
cell[3], celular
cell cycle, ciclo celular
cell division, división celular
cell membrane, membrana celular
cell respiration, respiración celular
cell specialization, especialización celular
cell theory, teoría celular
cell wall, pared celular
cellular respiration, respiración celular
cellulite, celulitis
celluloid, celuloide
cellulose, celulosa
Celsius (ºC), grado Celsius (ºC)
Celsius scale, escala Celsius
celtium, celtium
cement, cementar, cemento
cementation, cementación
Cenozoic Era, Era Cenozoica
center, centro
center of gravity, centro de gravedad
center of mass, centro de masas
centi, centi-, prefijo de centésima
centigrade, centígrado
centigram, centigramo
centiliter, centilitro
centimeter (cm), centímetro (cm)
centipede, ciempiés
central angle, ángulo central
central nervous system, sistema nervioso central
central processing unit (CPU), unidad central de procesamiento (CPU)
centrifugal force, fuerza

centrífuga
centrifuge, centrífuga, centrifugadora
centriole, centriolo
centripetal, centrípeto
centripetal force, fuerza centrípeta
centromere, centrómero
cephalic, cefálico
cephalochordate, cefalocordado
cephalopod, cefalópodo
cephalothorax, cefalotórax
cereal, cereal
cereal group, grupo de cereales
cerebellum, cerebelo
cerebral, cerebral
cerebral cortex, corteza cerebral
cerebral palsy, parálisis cerebral
cerebrospinal fluid, líquido cefalorraquídeo (LCR), cerebroespinal
cerebrum, telencéfalo
cerium, cerio
certain, certero, cierto
cervical, cervical
cervix, cuello uterino, cérvix uterino
cesium, cesio
cetacean, cetáceo
chain, cadena
chain reaction, reacción en cadena
chalcopyrite, calcopirita
chalk, creta
challenge, desafío
chamber, cámara
chameleon, camaleón
change[1], cambio
change of direction, cambio de dirección
change of motion, cambio de movimiento
change of speed, cambio de velocidad
change[2], cambiar
changing length of a shadow, cambio de longitud de una sombra
channel, canal, conducto
channel shape, forma del canal
chaos, caos
characteristic, característica
charcoal, carbón vegetal
charge, carga
charged, cargado
charging by conduction, carga por conducción
charging by induction, carga por inducción
chart, gráfico
cheek, mejilla
cheek bone, hueso malar
cheetah, guepardo
chelate, quelato
chemical, químico
chemical bond, enlace químico
chemical change, cambio químico
chemical composition, composición química
chemical energy, energía química
chemical engineering, ingeniería química
chemical equation, ecuación química
chemical name, nombre químico

chemical property, propiedad química
chemical reaction, reacción química
chemical system, sistema químico
chemical weathering, desgaste químico
chemisorption, quimisorción
chemistry, química
chemoautotroph, quimioautótrofos
chemosynthesis, quimiosíntesis
chemotherapy, quimioterapia
chemotroph, quimiotrofo
chert, sílex
chest, tórax, pecho
chickenpox, varicela
chimney, chimenea
chimpanzee, chimpancé
chin, barbilla, mentón
chip (computer), chip
chipmunk, ardilla
chiropractic, quiropráctica
chitin, quitina
chlorate, clorato
chloride, cloruro
chlorinate, clorar
chlorine, cloro
chlorine dioxide, dióxido de cloro
chlorite, clorito
chlorofluorocarbons, clorofluorocarbonos
chloroform, cloroformo
chlorohydrin, clorohidrina
chlorophyll, clorofila
chloroplast, cloroplasto
choice, elección
cholera, cólera
cholesterol, colesterol
choose, elegir
chord, cuerda
chordate, cordados
chorion, corión
chromatic, cromático
chromatic aberration, aberración cromática
chromatid, cromátida
chromatin, cromatina
chromatography, cromatografía
chrome green, cromo verde
chrome steel, acero cromado
chromic acid, ácido crómico
chromium, cromo
chromium sulfate, sulfato de cromo
chromosomal alteration, alteración cromosómica
chromosomal recombination, recombinación cromosómica
chromosome, cromosoma
chromosphere, cromósfera
chronological, cronológico
chronometer, cronómetro
chrysalis, pupa, crisálida
chyme, quimo
cicada, cicádido
cilia, cilios
ciliary motion, movimiento ciliar
cilium (pl. cilia), cilio
cinchona, quino
cinder cone, cono de ceniza
cinder cone volcano, volcán de cono de ceniza
cinnabar, cinabrio
circadian rhythm, ritmo circadiano
circle, círculo

circuit, circuito
  circuit board, placa de circuito
  circuit breaker, cortacircuitos
circular, circular
circulate, poner en circulación
circulation, circulación
circulatory, circulatorio
  circulatory system, aparato circulatorio
circumference, circunferencia
circumpolar stars, estrellas circumpolares
circumscribe, circunscribir
cirque, circo
cirrhosis, cirrosis
cirro, schirrus, cirrus
cirrocumulus, cirrocúmulo
cirrostratus, cirrostrato
cirrus, cirro, cirrus
  cirrus cloud, nube cirro
citrate, citrato
citric acid, ácido cítrico
citronella, citronella
citrus, cítrico
civil engineering, ingeniería civil
claim, reclamar
clam, almeja
clamp, abrazadera
clarinet, clarinete
clarity, claridad
class (biology), clase
classical mechanics, mecánica clásica
classical physics, física clásica
classification, clasificación
classify, clasificar
clastic rock, rocas clásticas
clavicle, clavícula
claw, garra, zarpa
clawed, armado de garras
clay, arcilla
cleavage, segmentación (biología), grieta (geología)
climate, clima
  climate variations, variaciones climáticas
climatology, climatología
climax community, comunidad clímax
climax fauna, fauna clímax
climax floras, flora clímax
clipping, recorte
clitoris, clítoris
cloaca, cloaca
clockwise, sentido horario
clone[1], clon
clone[2], clonar
cloning, clonación
closed circuit, circuito cerrado
closed circulatory system, sistema circulatorio cerrado
closed energy system, sistema cerrado de energía
clot, coágulo
cloud, nube
  cloud chamber, cámara de niebla
  cloud cover, cobertura de nubes
  cloud formation, formación de nubes
  cloud seeding, siembra de nubes
club, golpear
  club moss, Lycophyta

cnidarian, cnidarios, celentéreos
coacervate, coacervado
coagulate, coagular
coal, carbón
coarse, grueso
    coarse adjustment, ajuste grueso
coast, costa
coastal ocean, costa oceánica
coastline, línea costera
cobalamin, cobalamina
cobalt, cobalto
cobra, cobra
coccus, coco (bacteria)
coccyx, cóccix
cochlea, cóclea
cockle, cárdidos
cockroach, cucaracha
cocoon, capullo
code, código
codon, codón
coefficient, coeficiente
    coefficient of friction, coeficiente de fricción
    coefficient of linear expansion, coeficiente de expansión lineal
    coefficient of volume expansion, coeficiente de expansión volumétrica
coffee pods, sobres unidosis de café
cohesion, cohesión
cohesive, cohesivo
coil, bobina
coincidence, coincidencia
cold front, frente frío
collagen, colágeno
collate, cotejar
collating, clasificación
collect, reunir, juntar, recopilar
    collect information, recopilar información
collecting, recopilar, recolectar
    collecting duct, conducto colector
collection, colección
collide, chocar
colligative properties, propiedades coligativas
collinear, colineal
collision, colisión
    collision boundary, límite de colisión
colloid, coloide
colloidal dispersion, dispersión coloidal
colon, colon
colonization, colonización
colony, colonia
color, color
    color blind, daltónico
    color blindness, daltonismo
    color code, código de colores
    color scheme, combinación de colores
colorant, colorante
coloration, coloración
colorimetry, colorimetría
columbium, columbio
column, columna
coma, coma
comb jelly, ctenóforo
combat, combatir
combination, combinación
combine, combinar
combined gas law, ley

combinada de gas
combustion, combustión
combustion path, vía de combustión
comet, cometa
comfortable, confortable, cómodo
commercial, comercial
common, común
common cold, resfriado común
common denominator, denominador común
common divisor, común divisor
common factor, factor común
common good, (el) bien común
common ground, puntos en común, puntos de coincidencia
common ion effect, efecto del ion común
common logarithm, logaritmo común
common multiple, múltiplo común
commonalities, puntos en común
communicable, transmisible (enfermedad)
communication, comunicación
community, comunidad
community property, comunidad de bienes
commutator, conmutador
compact disk, disco compacto
compare, comparar
comparison, comparación
compartment, compartimento
compass, brújula
competing products, productos competitivos
competition, competencia
competitive exclusion principle, principio de exclusión competitiva
complement, complemento
complementary, complementario
complementary angles, ángulos complementarios
complementary color, color complementario
complementary pigment, pigmento complementario
complete protein, proteína completa
complex, complejo
complex carbohydrate, carbohidrato complejo
complex ion, ion complejo
complex multicellular, complejos multicelulares
complex number, número complejo
complexity, complejidad
component, componente
compose, componer
composite, compuesto
composite number, número compuesto
composite volcano, volcán compuesto
composition, composición
compost, abono, composta

compost pile, pila de compost, pila de composta
composting toilet, inodoro de composta
compound, compuesto
compound eye, ojo compuesto
compound leaf, hoja compuesta
compound lens, lente compuesta
compound microscope, microscopio compuesto
compounding, compuesto, mezcla
compress, comprimir
compressed gas, gas comprimido
compression, compresión
Compton effect, Efecto Compton
Computed Axial Tomography (CAT scan), tomografía axial computarizada (TAC)
computer, computadora
concave, cóncavo
concave lens, lente cóncava
concave mirror, espejo cóncavo
concentrated solution, solución concentrada
concentration, concentración
concept, concepto
conception, fecundación
conceptual definition, definición conceptual
concerning, referente
conch, caracola
conclude, concluir
conclusion, conclusión
concrete, hormigón
concurrent forces, fuerzas concurrentes
concussion, concusión
condensation, condensación
condensation polymer, polímeros de condensación
condensation reaction, reacción de condensación
condense, condensar
condition, condición
conditioned reflex, reflejo condicionado
conditioning, acondicionamiento
conduct, conducir
conductance, conductancia
conduction, conducción
conductivity, conductividad
conductor, conductor
cone, cono
configuration, configuración
configurational formula, fórmula configuracional
confine, confinar
conformation, conformación
congenital, congénito
conglomerate, conglomerado
Congo Red, Rojo Congo
congruent, congruente
conic projection, proyección cónica
conifer, conífera
coniferous, conífero
coniferous forest, bosque de coníferas
conjugate angles, ángulos conjugados
conjugate pair, par conjugado
conjugation, conjugación

conjunctiva, conjuntiva
connective tissue, tejido conectivo, tejido conjuntivo
consciousness, conciencia
consequences, consecuencias
conservation, conservación
  conservation factor, factor de conservación
  conservation law, ley de conservación
  conservation of change, conservación del cambio
  conservation of Earth resources, conservación de los recursos de la Tierra
  conservation of energy, conservación de energía
  conservation of environment, conservación del medio ambiente
  conservation of mass, conservación de la masa
conserve, conservar
considered, considerado
consist of, consiste en
consistent, consistente
consonance, consonancia
constant, constante
constellation, constelación
constipation, constipación
constituent, constituyente
construct, construir
construction, construcción
constructive, constructivo
consumer, consumidor
  consumer decision making, toma de decisiones del consumidor
  consumer product data, información sobre el consumo de productos
contact, contacto
  contact metamorphism, metamorfismo de contacto
contagion, contagio
contagious, contagioso
contain, contener
container[1], contenedor
container[2], recipiente
contaminant, contaminante
contaminate, contaminar
content, contenido
continent, continente
continental, continental
  continental climate, clima continental
  continental crust, corteza continental
  continental divide, divisoria continental, división continental
  continental drift, deriva continental
  continental glacier, glaciar continental
  continental line, línea continental
  continental plate, placa continental
  continental rise, emersión continental
  continental shelf, plataforma continental
  continental slope, talud continental
  continental tropical air mass, masa de aire tropical continental

continuous spectrum, espectro continuo
contour farming, agricultura de contorno
contour interval, intervalo de curvas de nivel
contour lines, curvas de nivel
contour map, mapa de contorno
contour ploughing, cultivo en contorno
contour plowing, cultivo en contorno
contraception, anticonceptivos
contraceptive, anticonceptivo
contract, contratar
contraction, contracción
contrast, contraste
contribute, contribuir
control, control
  control key (computer), tecla de servicio, tecla control
  control rod, barra de control
controlled experiment, experimento controlado
controlled variable, variable controlada
convalescence, convalecencia
convection, convección
convective circulation, circulación convectiva
convenient, conveniente
converge, converger
convergence, convergencia
  convergence zone, zona de convergencia
convergent, convergente
  convergent evolution, evolución convergente
  convergent plate boundary, límite de placas convergentes
converging lens, convergencia de la lente
conversion, conversión
convert, convertir
converter, convertidor
convex, convexo
  convex lens, lentes convexa
  convex mirror, espejo convexo
convulsion, convulsión
coordinate, coordenada
  coordinate system, sistema de coordenadas
coordinating system, sistema de coordinación
coordination, coordinación
copepod, copépodo
copolymer, copolímero
copper, cobre
coral, coral
  coral reef, arrecife de coral
cord, cuerda
cordillera, cordillera
core, centro, núcleo, esencia
Coriolis effect, efecto Coriolis
cork, corcho
corm, cormo
cornea, córnea
cornerstone, piedra angular
corolla, corola
corollary, corolario
corona, corona, halo
coronary, coronario
  coronary artery, arteria coronaria
  coronary circulation, circulación coronaria
corpus luteum, cuerpo lúteo

corpuscle, corpúsculo
correlation, correlación
corresponding, correspondiente
corrosion, corrosión
cortex, corteza
cortisone, cortisona
corundum, corindón
cosecant, cosecante
cosine, coseno
cosmic, cósmico
  cosmic background, fondo cósmico
  cosmic radiation, radiación cósmica
cosmology, cosmología
cosmopolites, cosmopolitas
cosmos, cosmos
cost accounting, contabilidad de costos
cost-benefit tradeoffs, costo-beneficio de las compensaciones
cotangent, cotangente
cotyledon, cotiledón
coulomb, culombio
Coulomb's Law, Ley de Coulomb
counterclockwise, sentido antihorario
countless, sinnúmero
course of action, curso de acción
covalence, covalencia
covalent, covalente
  covalent bond, enlace covalente
  covalent molecule, molécula covalente
cover, cubierta
  cover crop, cultivos de cobertura
coverslip, cubreobjetos
Cowper's gland, glándula de Cowper
cowrie, cauri
CPR, reanimación cardiopulmonar (RCP)
crack, agrietar
cracking, agrietamiento
craft, oficio
cranial nerve, nervio craneal, par craneal
crash, estrellar
crate, cajón
crater, cráter
crayfish, cangrejo de río
create, crear
crescent, media luna
  crescent moon, luna creciente
cresol, cresol
crest, cresta
Cretaceous, Cretáceo
cretinism, cretinismo
crevasse, grieta
cricket, grillo
crinoid, crinoideo
crisp, crujiente
criteria within constraints, criterios dentro de las limitaciones
critical, crítico
  critical angle, ángulo crítico
  critical mass, masa crítica
  critical point, punto crítico
  critical pressure, presión crítica
  critical temperature, temperatura crítica
crocodile, cocodrilo
crocodilian, cocodrilo
crop, cultivo, buche

crop rotation, rotación de cultivos
cross axis, eje transversal
crossbreed, cruzar
crossbreeding, cruzamiento
cross-bedding, estratificación cruzada
cross-cutting, transversales
cross-cutting relationships, relaciones intersectoriales
cross-fertilization, fertilización cruzada
crossing-over, sobrecruzamiento
cross-pollination, polinización cruzada
cross-section, sección transversal
crow, corneja
crowded, concurrido
crown, corona
crucible, crisol
crucible tongs, pinzas para tubo de ensayo
crust, corteza
crustacean, crustáceo
crustal down-warping, deformación cortical
cryogenics, criogenia
cryptic coloration, coloración críptica
crystal[1], cristal
crystal shape, forma de los cristales
crystal[2], cristalina
crystal lattice, red cristalina
crystal structure, estructura cristalina
crystalline, cristalino
crystalline structure, estructura cristalina
crystallization, cristalización
cube, cubo
cube root, raíz cúbica
cubic, cúbico
cubic centimeter, centímetro cúbico
cubic meter, metro cúbico
cultivated plant, planta cultivada
culture[1], cultivo
culture[2], cultivar
cumulonimbus, cumulonimbo
cumulus, cúmulo, cúmulus
cure, cura
curie, curio
curious, curioso
curium, curio
current, corriente
current theories, teorías actuales
currently, actualmente
curvature, curvatura
curve, curva
cutaneous, cutáneo
cuticle, cutícula
cutting, corte
cyan, cian
cyanide, cianuro
cyanogen, cianógeno
cybernetics, cibernética
cycle, ciclo
cyclic, cíclico
cyclic changes, cambios cíclicos
cyclic compound, compuesto cíclico
cycling, ciclismo
cycloalkane, cicloalcano
cyclohexane, ciclohexano
cyclone, ciclón

cyclonic system, sistema ciclónico
cyclosis, ciclosis
cyclotron, ciclotrón
cylinder, cilindro
cymbal, platillo
cyst, quiste
cysteine, cisteína
cystic fibrosis, fibrosis quística
cytochrome, citocromo
cytokinesis, citocinesis
cytology, citología
cytolysis, citólisis
cyton, Cyton
cytoplasm, citoplasma
cytoplasmic division, división citoplasmática
cytosine, citosina
cytoskeleton, citoesqueleto

daily cycle, ciclo diario
daily motion, movimiento diario
dam, dique, represa, embalse
damage, daño
dangerous, peligroso
dark reaction, reacciones de oscuridad
dart, dardo
Darwinism, Darwinismo
data, datos
  data auditing, auditoría de datos
  data collection sheet, hoja de recolección de datos
  data table, tabla de datos
daughter cell, célula hija
day, día
  day length, duración del día
DDT, DDT
De Broglie Principle, principio de De Broglie
deamination, desaminación
debate, debate
debris, desechos
deca-, deca-, prefijo de decámetro
decagon, decágono
decapod, decápodo
decay, decadencia, descomposición,
deceleration, desaceleración
deci-, deci-, de décimo
decibel, decibelio, decibel
deciduous, caducifolio
  deciduous tree, árbol de hoja caduca
decimal, decimal
  decimal fraction, fracción decimal
  decimal notation, notación decimal
  decimal place, lugar decimal
  decimal point, coma, punto decimal
  decimal system, sistema decimal
decimeter, decímetro
deck, cubierta
declination, declinación
decomposer, organismo saprofito
decomposition, descomposición
decrease, disminución
decrepitation, decrepitación
deduction, deducción

defecation, defecación
defective, defectuoso
defective gene, gen defectuoso
defend, defender
deficiency disease, enfermedad de deficiencia
deficient, deficiente
deficit, déficit
define, definir
definite, definitivo
definition, definición
deflect, desviar
defoliation, defoliación
deforestation, deforestación
deformation, deformación
deformation of rocks, deformación de las rocas
degeneration, degeneración
deglaciation, deshielo
degree, grado
degree of saturation, grado de saturación
dehydrating agent, agente deshidratante
dehydration, deshidratación
dehydration synthesis, síntesis por deshidratación
dehydrogenase, deshidrogenasa
dehydrogenation, deshidrogenación
deionized water, agua desionizada
deka-, deca-, prefijo de decámetro
dekameter, decámetro
delete key (computer), tecla suprimir
deletion, supresión
deliquescence, delicuescencia
deliver, entregar
delivery tube, tubo de entrega
delta, delta
demand, demandar
demographic transition, transición demográfica
demography, demografía
demonstrate, demostrar
demonstration, demostración
denaturation, desnaturalización
denatured alcohol, alcohol desnaturalizado
dendrite, dendrita
dendritic pattern, patrón dendrítico
dendrochronology, dendrocronología
denitrification, desnitrificación
denitrifying bacteria, bacterias desnitrificantes
denominator, denominador
density, densidad
density-dependant limiting factor, factor limitante dependiente de densidad
density-independent limiting factor, factor limitante independiente de densidad
dentin, dentina
dentistry, odontología
dentition, dentición
deoxyribonucleic acid (DNA), ácido desoxirribonucleico (ADN)
deoxyribose, desoxirribosa
depend, depender
dependent variable, variable dependiente
deplete, agotar

depletion, agotamiento
deposit, depósito, yacimiento (geología)
deposition, deposición
depositional basin, cuenca de depositación
depositional—erosional system, sistema de erosión y depósito
depressant, depresivo
depression, depresión
derivative, derivada (matemáticas)
derive, derivar
derived, derivado
  derived unit, unidad derivada
dermis, dermis
desalinization, desalinización, desalación
descend, descender
descending flow, flujo descendente
descent, descenso
describe, describir
desert, desierto
desertification, desertización
desiccate, secar
design[1], diseñar
design[2], diseño
desirable, deseable
destroy, destruir
destruction, destrucción
destructive, destructivo
  destructive interference, interferencia destructiva
detect, detectar
detection, detección
detector, detector
detergent, detergente
determine, determinar
detoxication, desintoxicación
detrimental, perjudicial
detritus, detrito
deuterium, deuterio
deuteron, deuterón
develop, desarrollar
developing agent, agente de desarrollo
development, desarrollo
deviate, desviarse
deviation, desviación
device, dispositivo
Devonian Period, Período Devónico
dew, rocío
  dew point, punto de rocío
dewclaw, espolón
dewdrop, gota de rocío
dewlap, papada
dextrorotatory, dextrógiro
dextrose, dextrosa, glucosa
diabetes, diabetes
diagnosis, diagnóstico
diagonal, diagonal
diagram, diagrama
  diagram key, diagrama indicador
dialysis, diálisis
diameter, diámetro
diamond, diamante, rombo
diaphragm, diafragma
diarrhea, diarrea
diastole, diástole
diastolic pressure, presión diastólica
diastrophism, diastrofismo
diatom, diatomea
diatomic gas, gas diatómico
diborane, diborano
dicarboxylic acid, ácido dicarboxílico
dichotomous, dicotómica

dichotomous key, clave dicotómica
dicotyledon, dicotiledóneas
died off, extinguirse
dielectric, dieléctrico
diesel engine, motor diesel
diet, dieta
diethylamine, dietilamina
differ, diferir
difference, diferencia
difference in electric potential, diferencia de potencial eléctrico
differentiation, diferenciación
diffract, difractar
diffraction, difracción
diffraction grating, rejilla de difracción
diffuse, difuso
diffuse reflection, reflexión difusa
diffusion, difusión
digest, digerir
digestion, digestión
digestive juice, jugo digestivo
digestive system, aparato digestivo
digestive tract, tubo digestivo
digit, dígito
digital, digital
dihybrid, dihíbrido
dihydroxy, dihidroxi
dike, dique
dilate, dilatar
dilation, dilatación
dilation of blood vessel, dilatación de los vasos sanguíneos
diluent, diluente
dilute, diluir
diluted solution, solución diluida
dilution, dilución
dimension, dimensión
dimer, dímero
dimethylketone, dimethylketone
dimmer, regulador
dimorphism, dimorfismo
dinitrobenzene, dinitrobenceno
dinosaur, dinosaurio
diode, diodo
dioecious, dioico
diolefin, diolefin
diorite, diorita
dioxide, dióxido
dip, sumergirse
dipeptide, dipéptido
diphtheria, difteria
dipole, dipolo
direct, directo
direct combination, combinación directa
direct combustion reaction, reacción de combustión directa
direct current, corriente continua
direct harvesting, recolección directa
direct rays, rayos directos
direction, dirección, sentido
directly, directamente
directory, directorio
directrix, directriz
disaccharide, disacárido
disadvantage, desventaja
disappearing trait, rasgo en vías de extinción
disaster, desastre
discharge, descarga

disconformity, disconformidad
discover, descubrir
discovery, descubrimiento
discs, discos
discuss, conversar, discutir
disease, enfermedad
disintegrate, desintegrar
disjunction, disyunción
disk, disco
  disk drive, unidad de disco
dislocation, dislocación
dispersal, dispersión
disperse, dispersar
dispersion, dispersión
displace, desplazar
displacement, desplazamiento
  displacement sediments, desplazamiento de sedimentos
  displacement series, desplazamiento de las series
disposal, eliminación
dispose, disponer
disproportionate, desproporcionada
disprove, refutar
disregard, indiferencia
disrupt, interrumpir
dissect, disecar, diseccionar
dissecting microscope, microscopio de disección
dissection, disección
dissipate, disipar
dissociation, disociación
dissolution, disolución
dissolve, disolver
dissonance, disonancia
distance, distancia
distant, distante
distillate, destilado
distillation, destilación
distinctive, distintivo
distort, distorsionar
distorted structure, estructura distorsionada
distortion, distorsión
distributary, distributario
distribute, distribuir
distribution, distribución
disturb, perturbar
disturbance, perturbación
diuretic, diurético
diurnal, diurno
dive, bucear
diver, buzo
diverge, divergir
divergence, divergencia
  divergence zone, zona de divergencia
divergent, divergente
  divergent plate boundary, límite de placas divergentes
diverging lens, lente divergente
diversity, diversidad
divide, dividir
dividend, dividendo
divider, divisor
division, división
divisor, divisor
DNA, ADN
  DNA fingerprinting, huellas de ADN
  DNA polymerase, ADN polimerasas
dock, muelle
dockhand, trabajador portuario, estibador
dodecagon, dodecágono
dodecahedron, dodecaedro
dog, perro

doldrums, zona de calmas ecuatoriales
dolomite, dolomita
dolphin, delfín
domain, dominio
dominance, dominio
dominant, dominante
  dominant gene, gen dominante
  dominant species, especie dominante
  dominant trait, rasgo dominante
donor, donante
dope, droga
Doppler, Doppler
  Doppler effect, Efecto Doppler
  Doppler shift, Efecto Doppler
dormancy, inactividad
dormant, aletargado, inactivo
dorsal, dorsal
dot diagram, diagrama de puntos
double, doble
  double bond, doble enlace
  double fertilization, doble fertilización
  double helix, doble hélice
  double slit diffraction, difracción de doble rendija
  double-pan balance, balanza de doble plato
down (feather), plumón
Down syndrome, síndrome de Down
downslope, pendiente
drag, arrastre
dragonfly, libélula
drainage, drenaje
  drainage basin, cuenca de drenaje
  drainage patterns, patrones de drenaje
draw, dibujar
drift, deriva
drilling, perforación
drink, trago
drone, zángano, abejón
drop, gota
droplet, gotita
dropper, gotero
drosophila, drosophila
drought, sequía
drug, droga
  drug abuse, uso indebido de drogas
  drug dependence, drogodependencia
  drug overdose, sobredosis
drum, tambor, bidón
drumlin, colina
drupe, drupa
dry, seco
  dry adiabatic lapse rate, gradiente adiabático seco
  dry bulb thermometer, termómetro de bulbo seco
  dry cell, pila seca
  dry ice, hielo seco
  dry land, tierra firme, tierra de secano
  dry measure, medida para áridos
duckweed, lenteja de agua
duct, conducto

ductile, dúctil
ductility, ductilidad
ductless gland, glándula de secreción interna
due, deber
dump, verter, descargar, basurero
dune, duna
duodecimal, duodecimal
duodenum, duódeno
duration of insolation, duración de la insolación
dust, polvo
dust storm, tormenta de polvo
duty, deber
dwarf, enano
dwarf planet, planeta enano
dwarf star, estrella enana
dwarfism, enanismo
dynamic, dinámico
dynamic geologic system, sistema geológico dinámico
dynamic process, proceso dinámico
dynamics, dinámica
dynamite, dinamita
dynamo, dinamo (dínamo)
dyne, dina
dysentery, disentería
dyslexia, dislexia
dysphasia, disfasia
dysprosium, disprosio

eagle, águila
ear, oído
ear canal, canal auditivo
eardrum, tímpano
early warning system, sistema de alerta temprana
Earth[1], Tierra
Earth Science Reference Tables, Tablas de Referencia para las Ciencias de la Tierra
Earth Systems, Sistemas terrestres
Earth's plates, placas de la Tierra
Earth's rotation, rotación de la Tierra
earth[2], tierra
earth science, ciencias de la tierra
earthquake, terremoto
earthquake epicenter, epicentro del terremoto
earthquake magnitude, magnitud del terremoto
earthworm, lombriz
earwig, tijereta
Eastern Hemisphere, hemisferio oriental
eat, comer
ebb (tide), reflujo
Ebola virus, virus de Ébola
eccentricity, excentricidad
echidna, equidna
echinoderm, equinodermo

echo, eco
eclipse, eclipse
eclipsing binary, binaria eclipsante
ecliptic, eclíptica
ecological, ecológica
    ecological niche, nicho ecológico
    ecological pyramid, pirámide ecológica
    ecological succession, sucesión ecológica
ecologically, ecológicamente
ecology, ecología
economic impacts, impactos económicos
economical, económico
economically, económicamente
economics, economía
ecosphere, ecosfera
ecosystem, ecosistema
    ecosystem diversity, diversidad de ecosistemas
ectoderm, ectodermo
eczema, eccema
eddy, remolino
edema, edema
edentate, desdentado
eel, anguila
effect, efecto
effective, efectivo
    effective collision, colisión efectiva
    effective resistance, resistencia efectiva
    effective work habits, hábitos de trabajo efectivos
effectively, efectivamente
effectiveness, eficacia
effector, efector
efferent, eferente
effervescence, efervescencia
efficiency, eficiencia
efficient, eficiente
efflorescence, eflorescencia
effort, esfuerzo
egestion, egestión
egg, huevo
El Niño, El Niño
    El Niño effect, efecto del Niño
Einstein, Einstein
einsteinium, einstenio
elastic, elástica
    elastic collision, choque elástico
elasticity, elasticidad
elbow, codo
    elbow joint, articulación del codo
electric, eléctrico
    electric cell, celda eléctrica
    electric current, corriente eléctrica
    electric field, campo eléctrico
    electric field intensity, intensidad del campo eléctrico
    electric field line, línea de campo eléctrico
    electric force, fuerza eléctrica
    electric generator, generador eléctrico
    electric potential, potencial eléctrico
electrical conductivity, conductividad eléctrica
electrical energy, energía

eléctrica
electricity, electricidad
electrocardiogram (ECG or EKG), electrocardiograma (ECG)
electrochemical, electroquímico
electrochemical cell, celda electroquímica
electrochemistry, electroquímica
electrode, electrodo
electrodeposition, electrodeposición
electrodynamics, electrodinámica
electroencephalogram, electroencefalograma
electrolysis, electrólisis
electrolyte, electrolito
electrolytic cell, celda electrolítica
electrolytic conduction, conducción electrolítica
electrolyze, electrolizar
electromagnet, electroimán
electromagnetic, electromagnético
electromagnetic force, fuerza electromagnética
electromagnetic induction, inducción electromagnética
electromagnetic radiation, radiación electromagnética
electromagnetic spectrum, espectro electromagnético
electromagnetic wave, onda electromagnética
electromagnetism, electromagnetismo
electromotive (EMF), electromotriz
electromotive force, fuerza electromotriz
electromotive series, serie electromotriz
electron, electrón
electron cloud, nube de electrones
electron microscope, microscopio electrónico
electronegative, electronegativo
electronic, electrónico
electronic balance, balanza electrónica
electronic communications network, red de comunicaciones electrónicas
electronic data processing, procesamiento electrónico de datos
electronic information, información electrónica
electronic materials, materiales electrónicos
electronic microscope, microscopio electrónico
electronic references, referencias electrónicas
electronic tube, válvula electrónica
electronic volt, electronvoltio
electrophoresis, electroforesis
electroplating, galvanoplastia

electropositive, electropositivo
electroscope, electroscopio
electrostatic force, fuerza electrostática
electrovalence, electrovalencia
electrovalent bonding, vinculación electrovalente
elementary particle, partícula elemental
elements, elementos
elephant, elefante
elephantiasis, elefantiasis
elevation, elevación
elevator, elevador
eliminate, eliminar
ellipse, elipse
elliptical, elíptico
 elliptical galaxies, galaxias elípticas
elm, olmo
elongation, elongación
 elongation region, region de elongación
 elongation zone, zona de elongación
email, correo electrónico
embed, integrar
embolism, embolia
embolus, émbolo
embryo, embrión
 embryo sac, saco embrionario
embryology, embriología
embryonic, embrionario
 embryonic membrane, membranas embrionarias
 embryonic stem cells, células madre embrionarias
emerald, esmeralda
emerge, surgir
emergence, emergencia, aparición
emergency, emergencia
 emergency action plan, plan de acción para emergencias
 emergency preparedness, preparación para emergencias
emigration, emigración
emission, emisión
 emission spectrum, espectro de emisión
emit, emitir
emphysema, enfisema
empirical, empírico
 empirical formula, fórmula empírica
empty, vacío
 empty set, conjunto vacío
emu, emú
emulsification, emulsificación
emulsion, emulsión
enable, permitir
enamel, esmalte
encephalitis, encefalitis
enclose, adjuntar
encounter, encontrarse con
end moraine, morrena final
endangered species, especie en vía de extinción
endemic, endémico
endocrine gland, glándula endocrina
endocrine system, sistema endocrino
endocrinology, endocrinología
endoderm, endodermo
endometrium, endometrio

endoparasite, endoparásitos
endorphin, endorfina
endoskeleton, endoesqueleto
endosperm, endospermo
endosperm nucleus, núcleo endospermo
endospore, endospora
endothermic, endotérmico
endpoint, extremo
energetic, energética
energetic state, estado energético
energy, energía
energy crisis, crisis energética
energy flow, flujo de energía
energy generation plan, plan de generación de energía
energy level, nivel de energía
energy pyramid, pirámide de energía
energy resource, fuente de energía
energy sink, disipador de energía
energy source, fuente de energía
engine, motor
engineer, ingeniero
engineering, ingeniería
engulf, sumergir
enlist, conseguir
enormous, enorme
enrich, enriquecer
enrichment, enriquecimiento
ensure, asegurar
enter, entrar
enter key (computer), tecla enter
enthalpy, entalpía
entire, todo
entomology, entomología
entrenched, atrincherado
entropy, entropía
environment, entorno, medio ambiente, ambiente
environmental, medio ambiente
environmental balance, equilibrio del medio ambiente
environmental changes, cambios en el medio ambiente
environmental consideration, consideración del medio ambiente
environmental impact, impacto ambiental
environmental impact statement, declaración de impacto ambiental
environmental implications, implicancias ambientales
enzyme, enzima
enzyme-substrate complex, complejo enzima-sustrato
eon, eón
epicenter, epicentro
epicotyl, epicótilo
epidemic, epidemia
epidemiology, epidemiología
epidermis, epidermis
epididymis, epidídimo
epiglottis, epiglotis
epilepsy, epilepsia
epinephrine, epinefrina

epiphyte, epifita
epithelial, epitelial
epithelium, epitelio
epoch, época
epoxy, epoxi
  epoxy resin, resina epoxi, resina epoxica
Epsom salts, sal de Epsom
equal, igual
equality of change, igualdad de cambio
equation, ecuación
equator, ecuador
equatorial plate, placa ecuatorial
equidistant, equidistante
equilibrant force, fuerza equilibrante
equilibrium, equilibrio
  equilibrium constant, constante de equilibrio
  equilibrium position, posición de equilibrio
equine, equino
equinox, equinoccio
equipment, equipo
equivalent, equivalente
  equivalent mass, masa equivalente
era, era
erbium, erbio
erect, erigir
  erect posture, postura erecta
erection, erección
erg, ergio
ergonomics, ergonomía
ergot, cornezuelo
Erlenmeyer flask, matraz de Erlenmeyer
erosion, erosión
  erosional - depositional system, erosión y depósito en sistemas
  erosional situation, situación de erosión
erratics, erráticas
error, error
  error extent, error de medida
erupt, entrar en erupción
escape velocity, velocidad de escape
escarpment, escarpa
Escherichia coli (E. coli), Escherichia coli (E. coli)
esker, esker
esophagus, esófago
essence, esencia
essential amino acid, aminoácido esencial
essentially, esencialmente
establish, establecer
establishment, establecimiento
ester, éster
esterification, esterificación
estimate, calcular
estimation, estimación, cálculo
estivate, estivar
estivation, estivación
estrogen, estrógeno
estrus, estro
estuary, estuario
ethanal, etanal
ethane, etano
ethanol, etanol
ethanolamine, etanolamina
ethene, eteno
ether, éter
ethical, ética
ethyl, etilo
  ethyl alcohol, alcohol etílico

ethyl ether, éter etílico
ethylamine, etilamina
ethylbenzene, etilbenceno
ethylene, etileno
ethylene bromide, bromuro de etileno
ethylene dichloride, dicloruro de etileno
ethylene glycol, etilenglicol
etiology, etiología
eucalyptus, eucalipto
euglena, euglena
eukaryote, eucariota
eukaryotic, eucariota
eukaryotic cell, célula eucariota
europium, europio
Eustachian tube, trompa de Eustaquio
eutectic, eutéctica
eutrophication, eutrofización
evaluate, evaluar
evaporate, evaporar
evaporation, evaporación
evaporites, evaporitas
evapotranspiration, evapotranspiración
even number, número par
evenly, igualmente
evenness, igualdad
event, evento
eventually, eventualmente
evergreen, perennifolio, siempreverde
evidence, evidenciar, evidencia
evolution, evolución
evolutionary, evolutivo
evolve, evolucionar
examination, examen
example, ejemplo
exception, excepción
excess, exceso
excessive, excesivo
excitation, excitación
excited state, estado de excitación
exclude, excluir
exclusive, exclusivo
excrete, excretar
excretion, excreción
excretory, excretor
exercise, ejercicio
exert, ejercer
exfoliation, exfoliación
exhalation, exhalación
exhaust, agotar
exhibit, exhibir
exist, existir
existence, existencia
exocrine gland, glándula exocrina
exon, exón
exoskeleton, dermatoesqueleto
exosphere, exósfera
exothermic, exotérmico
expand, expandir
expansion, expansión
expect, esperar
expectorant, expectorante
expel, expulsar
experiment[1], experimentar
experiment[2], experimento
experimental design, diseño experimental
experimenting, experimentar
expiration, expiración
explain, explicar
explanation, explicación
exploitation of fauna, explotación de la fauna
explore, explorar, investigar

explosion, explosión
explosive, explosivo
exponent, exponente
exponential, exponencial
  exponential function, función exponencial
  exponential growth, crecimiento exponencial
  exponential notation, notación exponencial
expose, exponer
exposed, expuesto
exposure, exposición
express[1], expresar
express[2], expreso
expressed, expresado
expression, expresión
extend, ampliar
extensor, extensor
exterior angle, ángulo exterior
external, externo
  external circuit, circuito exterior
  external fertilization, fertilización externa
  external force, fuerza externa
  external respiration, respiración externa
extinct[1], extinguir
extinct[2], extinguido
extinction, extinción
  extinction of fauna, extinción de la fauna
extinguishing agent, agente extintor
extracellular digestion, digestión extracelular
extract, extraer
extraction, extracción
extrapolate, extrapolar
extrapolation, extrapolación
extraterrestrial, extraterrestre
extreme, extremo
extremely, extremadamente
extrusion, extrusión
extrusive igneous, ígneas extrusivas
eye, ojo
eyebrow, ceja
eyelashes, pestañas
eyelid, párpado
eyepiece, ocular
eyespot, mancha ocular
eyestalk, pedúnculo ocular
eyrie, aguilera

facet, faceta
faceted, facetada
facilitated diffusion, difusión facilitada
fact, hecho
factice, facticio
factor, factor
factorial, factorial
facts and figures, datos y cifras
Fahrenheit (°F), grado Fahrenheit (°F)
fail, fallar
faint, desmayarse
fair test, evaluación justa
falcon, halcón
Fallopian tube, trompa de Falopio
fallout, polvillo radiactivo
family (biology), familia (biología)
fang, colmillo
farad, faradio
Faraday, Faraday

farsighted, hipermétrope
farsightedness, hipermetropía
fat[1], grasa
fat[2], grueso
fat[3], gordo
fatal, mortal
fathom, braza
fatigue, fatiga
fatty acid, ácido graso
fatty alcohol, alcohol graso
fault[1], falla
fault[2], defecto
fauna, fauna
feat, hazaña
feature, rasgo, característica
feces, heces
feedback, retroalimentación
    feedback mechanism, mecanismo de retroalimentación
feedstock, materia prima
feldspar, feldespato
feline, felino
felsic, félsica
female, hembra
    female gamete, gameto femenino
    female reproductive system, sistema reproductor femenino
femur, fémur
fermentation, fermentación
    fermentation tube, tubo de fermentación
fermium, fermio
fern, helecho
ferric, férrico, ferroso
ferroalloy, ferroaleaciones
ferrous, ferroso
    ferrous sulfate, sulfato ferroso
fertile, fértil
fertilization, fertilización
fertilizer, fertilizar
fetal, fetal
    fetal alcohol syndrome, síndrome de alcoholismo fetal
fetus, feto
fever, fiebre
fewer, menos
fiber, fibra
    fiber optics, fibra óptica
fiberglass, fibra de vidrio
Fibonacci sequence, sucesión de Fibonacci
fibrillation, fibrilación
fibrin, fibrina
fibrinogen, fibrinógeno
fibrous root system, sistema de raíces fibrosas
fibula, peroné
fiction, ficción
field, campo
    field manager, gerente de campo
    field of microscope, campo del microscopio
filament, filamento
file, archivo
filler, relleno
filter[1], filtro
filter[2], filtrar
filtered, filtrado
filtrate, filtrar
filtration, filtración
fin, aleta
final velocity, velocidad final
finches, pinzones
fine, fino
    fine adjustment, ajuste fino
finger, dedo
Finger Lakes, Lagos Finger
fingernail, uña

fingerprinting, toma de huellas dactilares
finite, finito
fiords, fiordos
fireball, estrella fugaz
firefly, lampírido, luciérnaga
firm, firme
first filial generation, primera generación filial
First Law of Motion, Primera Ley del Movimiento
First Law of Thermodynamics, Primera Ley de la Termodinámica
first neutron, neutrón primero
first quarter, primer trimestre
first-level consumer, consumidor de primer nivel
first-order line, línea de primer orden
firth, estuario
fish, pez, pescar
fission, fisión
  fission reactor, reactores de fisión
fissure, fisura
fixed, fijo
  fixed pulley, polea fija
  fixed star, estrella fija
fjord, fiordo
fl. oz. (fluid ounce), onza líquida
flagellate, flagelados
flagellum (pl. flagella), flagelo
flame, llama
flamingo, flamenco
flammable material, material inflamable
flare, llamarada
flash point, punto de inflamación
flask, matraz
flat, plano
flatfish, pleuronectiformes, peces planos
flatten, aplanar
flatworm, platelmintos, gusanos planos
flea, pulga
fledgling, polluelo
flexor, flexor
flick, película
flint, sílex
  flint glass, cristal de roca
flipper, aleta
float, flotar
flood, aluvión, inundación
  flood plain, llanura aluvial
  flood tide, marea creciente, pleamar
floppy disk, disquete, disco flexible
flora, flora
Florencia flask, matraz de Florencia
florescence, florescencia
flotation, flotación
flow, flujo
  flow chart, diagrama de flujo
  flow diagram, diagrama de flujo
  flow of energy, flujo de energía
flower, flor

fluid, fluido
fluid mosaic, mosaico fluido
fluke, trematodo, tipo de parásito
fluorescence, fluorescencia
fluorescent, fluorescente
fluorescent lamp, lámpara fluorescente
fluoridate, fluorizar
fluoridation, fluoración
fluoride, fluoruro, flúor
fluorine, flúor
fluorite, fluorita
fluorocarbon, fluorocarbono
fluoroscopy, fluoroscopia
fluorspar, fluorita
flute, acanalar
flux, flujo
fly (biology), mosca
foam, espuma
focal, focal
focal length, distancia focal
focal point, punto focal
foci, focos
focus (pl. foci), foco
fog, niebla
fold, pliegue
folded, plegado
folded strata, estratos plegados
foliated, foliada
folic acid, ácido fólico
follicle, folículo
follicle stimulating hormone (FSH), hormona folículo estimulante (FSH)
food, alimento
food allergen, alimentos alergénicos
food chain, cadena alimenticia
food poisoning, intoxicación alimentaria
food pyramid, pirámide alimenticia
food vacuole, vacuolas alimenticias
food web, red alimenticia
foot (pl. feet), pie, pata
foot-and-mouth disease, glosopeda, fiebre aftosa del ganado
force, fuerza
force of friction, fuerza de fricción
force of gravity, fuerza de gravedad
forceps, fórceps
forebrain, prosencéfalo, cerebro anterior
forecast, pronóstico
forehead, frente
forensic chemistry, química forense
forensic science, medicina forense
forest[1], bosque
forest[2], forestal
forest conservation, conservación forestal
fork, tenedor
form[1], formar
form[2], forma, formulario
formaldehyde, formaldehído, metanal
formation, formación
former, anterior
formic acid, ácido fórmico
formula, fórmula
formula mass, fórmula de la masa

fossil, fósil
fossil evidence, evidencia fósil
fossil fuel, combustible fósil
fossilization, fosilización
Foucault pendulum, péndulo de Foucault
foundation, creación, fundamento
fowl, ave de corral
fox, zorro
fractal, fractal
fraction, fracción
fractional distillation, destilación fraccionada
fractionation, fraccionamiento
fracture, fractura
fragment, fragmento
frame[1], marco
frame of reference, marco de referencia
frame[2], enmarcar
framework, armazón, marco
francium, francio
fraternal twin, hermano gemelo
free energy, energía libre
free fall, caída libre
free radical, radical libre
freedom, libertad
free-living virus, virus libre
freeze, congelar
freeze drying, liofilización
freezing, congelación
freezing point, punto de congelamiento
freezing point depression, depresión del punto de congelación
Freon, Freón
frequency, frecuencia
freshwater, agua dulce
friction, fricción
frictional, friccional
frictional drag, fricción de arrastre
frictionless, sin fricción
frog, rana
frond, fronda
front, frente
frontal, frontal
frontal bone, hueso frontal
frontal boundary, límite frontal
frontal lobe, lóbulo frontal
frost, escarcha, helada
frostbite, lesión por congelación
frozen, congelado
fructose, fructosa
fruit, fruta
fruit fly, mosca de la fruta
fuel, combustible
fuel cell, celda de combustible
fulcrum, fulcro
full moon, luna llena
fumarole, fumarola
fume, humo
function[1], funcionar
function[2], función, funcionamiento
functional group, grupo funcional
fundamental, fundamental
fundamental particle, partícula fundamental
fungi, hongos
fungus, hongo

funnel, embudo
fur, piel
furnace, horno
furthermore, además
fuse, fusible
fused salt, sal fundida
fusible alloy, aleación fusible
fusion, fusión
  fusion reactor, reactor de fusión

gabbro, gabro
gadolinium, gadolinio
gain, adquirir, ganar
galactose, galactosa
galactosemia, galactosemia
galaxy, galaxia
galena, galena
gall, agalla
gallbladder, vesícula biliar
gallium, galio
  gallium arsenide, arseniuro de galio
gallon, galón
gallstone, cálculos biliares
galvanic, galvánico
  galvanic cell, pila galvánica, celda galvánica
galvanization, galvanización
galvanometer, galvanómetro
gamet, granate
gamete, gameto
gametogenesis, gametogénesis
gametophyte, gametofito
gamma globulin, gama globulina
gamma ray, rayo gama
ganglion, ganglio
gap, brecha
garbage, basura
garnet, garnate
gas, gas
  gas exchange, intercambio de gases
  gas giants, planetas gaseosos gigantes
  gas phase, fase del gas
gaseous exchange, intercambio gaseoso
gasification, gasificación
gasket, empaque, empacadora, empaquetadura, junta
gasoline, gasolina
gastric, gástrico
  gastric juice, jugo gástrico
gastrointestinal tract, sistema gastrointestinal
gastropod, gasterópodo
gastrovascular cavity, cavidad gastrovascular
gastrovascular glycerol, glicerol gastrovascular
gastrula, gástrula
gastrulation, gastrulación
gathering and processing information, recolección y procesamiento de información
gauge, calibre
gear, engranaje
Geiger-Müller tube, tubo Geiger Müller
gel, gel
  gel electrophoresis, electroforesis en gel
gelatin, gelatina

gem, gema
gene, gen
gene frequency, frecuencia genética
gene linkage, ligación genética
gene pool, reserva genética
gene therapy, terapia génica
general theory of relativity, teoría general de la relatividad
generalization, generalización
generate, generar, producir
generation, generación
generator, generador
genetic, genético
genetic code, código genético
genetic counseling, asesoramiento genético
genetic disease, enfermedad genética
genetic diversity, diversidad genética
genetic engineering, ingeniería genética
genetic fingerprinting, identificación genética
genetic map, mapa genético
genetic marker, marcador genético
genetic material, material genético
genetic variation, variación genética
genetically, genéticamente
genetics, genética
genitals, genitales
genome, genoma
genotype, genotipo
genus, género
geocentric, geocéntrico
geocentric model, modelo geocéntrico
geochemistry, geoquímica
geode, geoda
geodesic dome, cúpula geodésica
geographic isolation, aislamiento geográfico
geographic poles, polos geográficos
geography, geografía
geologic hazard, peligros geológicos
geologic time, edad geológica
geologist, geólogo
geology, geología
geomagnetism, geomagnetismo
geometric, geométrico
geometric isomer, isómero geométrico
geometry, geometría
geophysics, geofísica
geosyncline, geosinclinal
geothermal, geotérmica
geothermal energy, energía geotérmica
geothermal heat, calor geotérmico
germ, germen
germ cell, célula germen
germ theory of disease, teoría de los gérmenes de la enfermedad
German measles, rubéola
germanium, germanio
germinate, germinar
germination, germinación
gestation, gestación
gestation period,

período de gestación
geyser, géiser
G-force, fuerza G
GH (growth hormone), hormona de crecimiento
giant star, estrella gigante
giantism, gigantismo
gibberellin, giberelina
gibbous, gibosa (luna)
giga, giga
gigabyte, gigabyte
gill, branquia
gingiva, encía
ginkgo, ginkgo
give, dar
gizzard, molleja
glacial action, acción glaciar
glacial lakes, lagos glaciares
glacier, glaciar
glance, mirada
gland, glándula
glass, vidrio, vaso
    glass electrode, electrodo de vidrio
glassware, cristalería
glaucoma, glaucoma
glider, planeador
global, global
    global climate, clima global
    global distribution, distribución global
    Global Positioning System (GPS), Sistema de Posicionamiento Global (GPS)
    global scale, escala global
    global warming, calentamiento global
globulin, globulina
glomerulus, glomérulo
gloss meter, brillómetro
glucagon, glucagón
glucose, glucosa
glue, pegamento
gluon, gluones
glutamine, glutamina
glyceride, glicérido
glycerin, glicerina
glycerol, glicerol
glycine, glicina
glycogen, glucógeno
glycol, glicol
glycolysis, glucólisis
gnat, mosquito
gneiss, gneis
goat, cabra, chiva
goggles, gafas de protección
goiter, bocio
gold, oro
gonad, gónada
gonadotropin, gonadotropina
gonorrhea, gonorrea
googol number, número googol
gorge, cañón, desfiladero
gorilla, gorila
gout, gota
govern, gobernar
graded bedding, sedimentación ordenada
gradient, gradiente
gradual, gradual
gradualism, gradualismo
gradually, gradualmente
graduated beaker, vaso de precipitado graduado
graduated cylinder, probeta
graduated pipette, pipeta graduada
graft, injerto
grafting, injerto
grain, grano

gram (g), gramo (g)
gram atomic mass, gramo de masa atómica
gram equivalent mass, gramo de masa equivalente
gram molecular mass, gramo de masa molecular
gram-molecular volume, volumen gramo-molecular
grana, grana
granite, granito
graph, gráfico
graphic, gráfico
graphical, gráfico
graphics, gráficos
graphite, grafito
grass, hierba, césped
grasshopper, saltamontes
grassland, pradera
grating, rejilla
gravel, grava
gravimeter, gravímetro
gravimetric analysis, análisis gravimétrico
gravitational, gravitatorio
gravitational attraction, atracción gravitacional
gravitational constant, constante gravitacional
gravitational field, campo gravitacional
gravitational force, fuerza gravitacional
gravitational mass, masa gravitacional
graviton, gravitación
gravity, gravedad
gray (Gy), gray (Gy)
gray matter, materia gris
grease, grasa
great circle, círculo mayor
greatest, el mayor, mejor, más importante
green, verde
green algae, algas verdes
green plant, planta verde
green revolution, revolución verde
greenhouse effect, efecto invernadero
greenhouse gas, gas de efecto invernadero
Greenwich Mean Time, tiempo medio de Greenwich
grid, red
groins, ingle
grooves, surcos
ground, tierra
ground moraine, suelo de morrena
ground state, estado fundamental
ground-sill, solera, viga de carrera
grounding, conexión a tierra
groundwater, agua subterránea
group[1], agrupar
group[2], grupo
grouping, agrupamiento
grow, crecer
growth, crecimiento
growth hormone (GH), hormona de crecimiento
growth ring (biology), anillo de crecimiento
guanine, guanina

guard cell, célula oclusiva
gulf, golfo
Gulf Stream, (la) corriente del Golfo
gullet, garganta, esófago
gully, barranco, desfiladero
gulp, trago
gum, encía
gun, pistola
guncotton, algodón pólvora
gut, intestino
guyot, guyot, monte submarino
gymnosperm, gimnospermas
gynecology, ginecología
gypsum, yeso
gypsy moth, lagarta peluda, gata peluda
gyroscope, giroscopio

habit, hábito
habitat, hábitat
habitat fragmentation, fragmentación del hábitat
hackle, pelaje
hafnium, hafnio
hail, granizo
hair, pelo
hair follicle, folículo piloso
half, media
half-life, vida media
half-reaction, semirreacción
halite, hálito
hallucinogen, alucinógeno
halo, halo
halogen, halógeno
halt, detener
hammer, martillo
hamstring, tendón de la corva
hand, mano
hand lens, lupa de mano
handle, manipular, manejar
hanging valley, valle colgante
hanging wall, pared inclinada
hard, duro(a)
hard disk, disco duro
hard drive, unidad de disco duro
hard water, agua dura
hardness, dureza
hardware, hardware
Hardy-Weinberg Law, ley de Hardy-Weinberg
hare, liebre
harm, dañar
harmful, nocivo, dañino
harmless, inofensivo
harmonic, armónico
harmonious, armonioso
harvesting, cosecha
hatching, incubación
haul, acarrear, arrastrar
Haversian canal, canal Haversiano
hawk, halcón
hay, heno
hay fever, fiebre del heno
hazard, peligro
hazardous waste, residuo peligroso
head, cabeza

head erosion, erosión de la línea lateral y la cabeza

heading angles, ángulo de incidencia

headlamp, faro

headland, cabo

health, salud

healthy habit, hábito saludable

hear, oír, escuchar

hearing, audiencia

heart, corazón

heart attack, ataque cardíaco

heart muscle, músculo cardíaco

heart transplant, transplante de corazón

heartbeat cycle, cíclo cardíaco, frecuencia cardíaca

heartburn, ardor de estómago, acidez, pirosis

heart-lung machine, máquina corazón-pulmón

heat[1], calentar

heat[2], calor

heat effect, efecto del calor

heat exchange, transferencia de calor

heat exhaustion, colapso por exceso de calor

heat loss, pérdida de calor

heat of combustion, calor de combustión

heat of condensation, calor de condensación

heat of crystallization, calor de cristalización

heat of dilution, calor de dilución

heat of formation, calor de formación

heat of fusion, entalpía de fusión, calor de fusión

heat of hydration, calor de hidratación

heat of reaction, calor de reacción

heat of solution, entalpía de solución

heat of sublimation, calor de sublimación

heat of transition, calor de transición

heat of vaporization, calor de vaporización

heat pump, bomba de calor

heat transfer, transferencia de calor

heat[3], térmico

heat budget, balance térmico

heat energy, energía térmica

heat engine, motor térmico

heat rash, sarpullido

heat treatment, termoterapia

heavy, pesado, fuerte

heavy hydrogen, hidrógeno pesado

heavy metal, metal pesado

heavy water, agua pesada

hectometer, hectómetro
heel, talón
height, altura
Heisenberg uncertainty principle, principio de incertidumbre de Heisenberg
helicopter, helicóptero
heliocentric, heliocéntrico
 heliocentric model, modelo heliocéntrico
heliotropism, heliotropismo
helium, helio
helix, hélice
hematite, hematita
hematoma, hematoma
hemiplegia, hemiplejia
hemisphere, hemisferio
hemlock, cicuta
hemoglobin, hemoglobina
hemolysis, hemólisis
hemophilia, hemofilia
hemorrhage, hemorragia
hence, por lo tanto
heparin, heparina
hepatic, hepático
 hepatic artery, arteria hepática
 hepatic portal circulation, circulación portal hepática
hepatitis, hepatitis
heptagon, heptágono
herb, hierba
herbicide, herbicida
herbivore, herbívoro
hereditary, hereditario
heredity, herencia
heritable, heredable
hermaphrodite, hermafrodita
hernia, hernia, ruptura
herpes, herpes
herpetology, herpetología
hertz, hertz
heterocyclic, heterocíclico
heterogeneous reaction, reacción heterogénea
heterotroph hypothesis, hipótesis heterótrofa
heterotrophic, heterótrofo
 heterotrophic nutrition, nutrición heterotrófica
heterozygous, heterocigoto
 heterozygous genotype, genotipo heterocigótico
hexagon, hexágono
hexagonal, hexagonal
hiatus, interrupción
hibernation, hibernación
hiccup, hipo
high, alto, fuerte, agudo
 high blood pressure, hipertensión arterial
 high energy bond, enlace de alta energía
 high explosive, explosivo de gran potencia
 high noon, mediodía
 high point, momento culminante
 high polymer, polímero alto
 high pressure, presión alta
 high temperature source, fuente de alta temperatura
 high tide, marea alta
 high winds, viento fuerte

hiker, caminante, excursionista
hill, colina, cerro
hillslope, ladera
hilly, montañoso
hilum, hilio
hindbrain, rombencéfalo
hip, cadera
  hip joint, articulación de la cadera
hipbone, hueso coxal
hippopotamus, hipopótamo
histamine, histamina
histidine, histidina
histology, histología
histone, histona
HIV (human immunodeficiency virus), VIH (virus de la inmunodeficiencia humana)
hive, colmena
hives, urticaria
hockey, hockey
hollow, excavar, vaciar
holmium, holmio
hologram, holograma
holography, holografía
home key (computer), tecla inicio
homeopathy, homeopatía
homeostasis, homeostasis
homeowner, dueño de casa
hominid, homínido
Homo sapiens, Homo sapiens
homocyclic, homocíclico
homogeneous reaction, reacción homogénea
homogenization, homogeneización
homologous, homólogo
  homologous series, series homólogas
homolosine projection, proyección homolosena
homopolymer, homopolímero
homozygous, homocigoto
hoofed animal, animales con pezuñas
hook, enganchar, pescar
Hooke's Law, ley de elasticidad de Hooke
hookworm, anquilostoma
horizon, horizonte
horizontal, horizontal
  horizontal sorting, clasificación horizontal
horizontally, horizontalmente
hormone, hormona
horn, cuerno
hornblende, hornablenda
horology, horología
horse, caballo
horsepower, caballo de vapor
horticulture, horticultura
host, huésped
hot, caliente
  hot spot, zona caliente
  hot spring, aguas termales
  hot water bath, baño de agua caliente
hot-melt, fusión en caliente
Hubble's Law, Ley de Hubble
hue, tono, matiz
huge, enorme
hull, vaina, cáscara
human, humano
  human activities, actividades humanas
  human immunodeficiency virus (HIV), virus de la inmunodeficiencia humana (VIH)

humectant, humectante
humerus, húmero
humid, húmedo
  humid climate, clima húmedo
humidity, humedad
hummingbird, colibrí
humoral immunity, inmunidad humoral
humus, humus
hunting, caza
hurricane, huracán
husbandry, explotación agrícola, explotación ganadera
husk, vaina, cáscara
hybrid, híbrido
  hybrid vigor, vigor híbrido, vigor heterosis
hybridization, hibridación
hydra, hidra
hydrate, hidrato
hydration, hidratación
hydraulic, hidráulico
  hydraulic pump, bomba hidráulica
  hydraulic system, sistema hidráulico
hydraulics, hidráulica
hydride, hidruro
hydrocarbon, hidrocarburo
hydrochloric acid, ácido clorhídrico
hydrochloride, clorhidrato
hydrocolloid, hidrocoloide
hydrocyanic, cianhídrico
  hydrocyanic acid, ácido cianhídrico
hydrodynamics, hidrodinámica
hydroelectric power, energía hidroeléctrica
hydrogen, hidrógeno
  hydrogen acceptor, aceptor de hidrógeno
  hydrogen bomb, bomba de hidrógeno
  hydrogen bond, enlace de hidrógeno
  hydrogen chloride, cloruro de hidrógeno
  hydrogen cyanide, ácido cianhídrico, cianuro de hidrógeno
  hydrogen fluoride, fluoruro de hidrógeno
  hydrogen iodide, yoduro de hidrógeno
  hydrogen peroxide, peróxido de hidrógeno
  hydrogen sulfide, ácido sulfhídrico, sulfuro de hidrógeno
hydrogenation, hidrogenación
hydrogenolysis, hidrogenolisis
hydrology, hidrología
hydrolysis, hidrólisis
  hydrolysis constant, constante de hidrólisis
hydrometer, hidrómetro
hydronium ion, ion hidronio
hydrophilic, hidrofílico
hydrophobic, hidrofóbico
hydroponics, hidroponía
hydrosphere, hidrósfera
hydrostatics, hidrostática
hydrothermal, hidrotermal
hydrotropism, hidrotropismo
hydrous, hidratado
hydroxyl, oxidrilo, hidroxilo
  hydroxyl group, grupo hidroxilo

hyena, hiena
hygrometer, higrómetro
hygrophyte, higrofita
hygroscopic, higroscópico
hymen, himen
hyperactivity, hiperactividad
hyperbola, hipérbola
hyperlink, hipervínculo
hyperparasitism, hiperparasitismo
hypertension, hipertensión
hypertext, hipertexto
hyperthyroidism, hipertiroidismo
hypnosis, hipnosis
hypochlorous acid, ácido hipocloroso
hypochondria, hipocondría
hypocotyl, hipocotilo
hypodermic needle, aguja hipodérmica
hypoglycemia, hipoglucemia
hyposecretion, hiposecreción
hypotenuse, hipotenusa
hypothalamus, hipotálamo
hypothermia, hipotermia
hypothesis, hipótesis
hypothetical, hipotético
hypothyroidism, hipotiroidismo

ibuprofen, ibuprofeno
ice, hielo
  ice age, glaciación
  ice cube, cubito de hielo
  ice sheet, capa de hielo
iceberg, iceberg
icecap, manto glaciar
ICF (intercellular fluid), LIC (líquido intracelular)
ichthyology, ictiología
ichthyosaur, ictiosauro
ideal, ideal
  ideal gas, gas ideal
  ideal gas law, ley del gas ideal
  ideal mechanical advantage (IMA), ventaja mecánica ideal (IMA)
identical, idéntico
  identical twin, gemelo idéntico
identify, identificar
  identify patterns, identificar los patrones
identity, identidad
  identity period, período de identidad
igneous, ígneo
  igneous rock, roca ígnea
  igneous rock formation, formación de roca ígnea
ignite, encender
ignition, encendido
iguana, iguana
ileum, íleon
illuminance, iluminancia
illuminate, iluminar
illuminated body, cuerpo iluminado
illumination, iluminación
illusion, ilusión
illustrate, ilustrar
image, imagen
imaginary, imaginario
  imaginary number,

número imaginario
imagine, imaginar
imago, imago
immerse, sumergir
immersion, inmersión
immersion heater, calentador de inmersión
immigration, inmigración
immiscible, inmiscible
immune, inmune
immune response, respuesta inmune
immune system, sistema inmunológico, sistema inmunitario
immunity, inmunidad
immunization, inmunización
immunodeficiency, inmunodeficiencia
immunogenicity, inmunogenicidad
immunology, inmunología
impact[1], impactar
impact[2], impacto
impact crater, cráter de impacto, astroblema
impact of information systems, impacto de los sistemas de información
impact[3], impactante
impact event, suceso impactante
impart, impartir
impedance, impedancia
impermeable, impermeable
impetigo, impétigo
impingement black, negro de carbón
implantation, implantación
implication, implicancia
import, importar
importation, importación
imported cases, casos importados
impossible, imposible
impregnation, impregnación
imprint, imprimir
impulse, impulso
impurity, impureza
in vitro, in vitro
in vitro fertilization, fertilización in vitro
inappropriate, inapropiado
inborn immunity, inmunidad innata
inbreed, endogamia
inbreeding, endogamia
incandescent, incandescente
incandescent lamp, lámpara incandescente
inch, pulgada
incidence, incidencia
incident, incidente
incident insolation, insolación incidente
incident pulse, pulso incidente
incident wave, onda incidente
incision, incisión
incisor, incisivo
inclined, inclinado
inclined plane, plano inclinado
include, incluir
inclusion, inclusion
incomplete protein, proteína incompleta
incompressible, incompresible
increase[1], aumentar
increase[2], aumento
increment, incremento

incubate, incubar
incubation period, período de incubación
Indanthrene Blue, Indanthrene Azul
independence, independencia
independent assortment, distribución independiente
independent variable, variable independiente
index, índice
index fossil, fósil índice
index of refraction, índice de refracción
index finger, dedo índice
Indian Red, tipo de suelo rojo
indicate, indicar
indication, indicación
indicator, indicador
indigenous, indígena
indigo, añil
indirect, indirecto
indium, indio (elemento químico)
individual[1]**,** individuo
individual[2]**,** individual
induction, inducción
induction coil, bobina de inducción, bobina de Ruhmkorff
inductive, inductivo
inductive reactance, reactancia inductiva
industrial alcohol, alcohol industrial
industrial diamonds, diamantes industriales
industrial hazard, peligro industrial
industrial melanism, melanismo industrial
industrialization, industrialización
inelastic, inflexible
inelastic collision, colisión inelástica
inert, inerte
inertia, inercia
inertial mass, masa inerte
inexpensive, económico
infantile paralysis, parálisis infantil
infection, infección
infectious, infeccioso
infer, inferir
inference, inferencia
infertile, estéril
infiltration, infiltración
infinite, infinito
infinity, infinito
inflammable, inflamable
inflammation, inflamación
inflammatory response, respuesta inflamatoria
inflate, inflar
inflorescence, inflorescencia
influenza, influenza
information, información
information technology, tecnología de la información
informed decisions, decisiones informadas
infrared, infrarrojo
infrared spectroscopy, espectroscopia infrarroja
infrasound, infrasonido
ingestion, ingestión
ingredient, ingrediente
inhalation, inhalación
inherit, heredar

inheritance, herencia
inherited adaptation, adaptación heredada
inherited trait, rasgo hereditario
inhibition, inhibición
  inhibition center, centro de inhibición
inhibitor, inhibidor
initial, inicial
  initial momentum, momento inicial
  initial velocity, velocidad inicial
initiating explosive, explosivo inicial
ink, tinta
inner core, núcleo interno
inner ear, oído interno
inoculation, inoculación
inorganic, inorgánico
  inorganic analysis, análisis inorgánico
  inorganic chemistry, química inorgánica
  inorganic compound, compuesto inorgánico
  inorganic solid element, elemento sólido inorgánico
input, entrada
inquiry into phenomena, investigación sobre los fenómenos
insect, insecto
insecticide, insecticida
insectivore, insectívoro
insectivorous plant, planta insectívora
insert, insertar
insolation, insolación
insoluble, insoluble
inspect, inspeccionar
inspiration, inspiración
instant, instante
instantaneous, instantáneo
  instantaneous speed, velocidad instantánea
  instantaneous velocity, velocidad instantánea
instantaneously, instantáneamente
instead, en vez de
instinct, instinto
instrument, instrumento
instrumental analysis, análisis instrumental
insulate, aislar
insulator, aislante
insulin, insulina
intake, ingesta
integer, número entero
integral, integral
integrate, integrar
integrated circuit, circuito integrado
integration, integración
integument, integumento
intensity, intensidad
  intensity of insolation, intensidad de la insolación
  intensity of radiation, intensidad de la radiación
interact, interactuar
interaction, interacción
interbreeding, cruzamiento
intercellular, intercelular
  intercellular fluid (ICF), líquido intracelular (LIC)
intercept, interceptar
intercourse, coito
interdisciplinary problems, problemas de carácter interdisciplinario

interface, interfase
interfere, interferir
interference, interferencia
interferon, interferón
intergrowth, intercrecimiento
interionic attraction, atracción interiónica
interior, interior
  interior angle, ángulo interior
interlock, entrelazar
intermediate, intermedio
intermetallic compound, compuesto intermetálico
internal, interno
  internal circuit, circuito interno
  internal development, desarrollo interno
  internal energy, energía interna
  internal fertilization, fertilización interna
  internal force, fuerza interna
  internal medicine, medicina interna
internally, internamente
International Date Line, línea internacional de cambio de fecha
international unit, unidad internacional
Internet, Internet
interneuron, interneurona
interparticle, interpartícula
interphase, interfase
interpolate, interpolar
interpret, interpretar
  interpret data, interpretar los datos
interpretation, interpretación
interrelate, interrelacionar
interrelationship, interrelación
intersection, intersección
interspecific competition, competencia interespecífica
interstage, interfase
interstellar, interestelar
interstitial, intersticial
interval, intervalo
intestinal juice, jugo intestinal
intestine, intestino
intracellular digestion, digestión intracelular
intravenous, intravenoso
introduce, introducir
intron, intrón
intrusion, intrusión
intrusive igneous rock, rocas ígneas intrusivas
invasion, invasión
invasive species, especie invasiva
invent, inventar
inverse, inverso
  inverse variation, variación inversa
  inverse-square law, ley del inverso del cuadrado
inversion, inversión
invert, invertir
invertebrate, invertebrado
investigation, investigación
involuntary, involuntario
  involuntary muscle, músculo involuntario
involve, involucrar
inward, interno
iodide, yoduro

iodine, yodo
ion, ión
ionic bond, enlace iónico
ionic bonding, enlace iónico
ionic conduction, conducción iónica
ionization, ionización
ionization constant, constante de ionización
ionization energy, energía de ionización
ionization potential, potencial de ionización
ionize, ionizar
ionogen, ionógeno
ionosphere, ionósfera
ion-product, producto iónico
ion-product constant of water, constante del producto iónico del agua
iridium, iridio
iris, iris
iron (Fe), hierro (Fe)
iron blue, color gris azulado
iron oxide, óxido de hierro
irradiate, irradiar
irrational number, número irracional
irregular, irregular
irregular galaxies, galaxias irregulares
irreversible, irreversible
irrigate, irrigar
irritability, irrigabilidad
island, isla
islet, islote
islets of Langerhans, islotes de Langerhans
isobar, isobara
isolate, aislar
isolated, aislado
isolated system, sistema aislado
isolation, aislamiento
isoline, línea de nivel, curva de nivel
isoline map, mapa de isolíneas
isomer, isómero
isomerization, isomerización
isopropyl alcohol, alcohol isopropílico
isosceles, triángulo isósceles
isostasy, isostasia
isosurface, isosuperficie
isotactic, isotáctico
isotherm, isotermo
isotope, isótopo
IUPAC system, Sistema IUPAC
ivory, marfil
ivory black, negro de marfil

J

jade, jade
jasper, jaspe
jaundice, ictericia
jaw, mandíbula, maxilar
jaw bone, quijada
jawless fish, agnatos
jejunum, yeyuno
jellyfish, medusa
jet, chorro
jet plane, avión a reacción
jet propulsion, propulsión a chorro
jet stream, corriente en chorro

jetty, embarcadero
jeweler, joyero
joint, unión, articulación, en conjunto
joint ventures, transacciones conjuntas, operaciones conjuntas
joule, julio, joule
journal, periódico
jungle, selva, jungla
Jupiter, Júpiter
juvenile, juvenil
juvenile hormone, hormona juvenil

kame, kame
kangaroo, canguro
kaolin, caolín
karst topography, topografía kársticas
karyotyping, cariotipo
keg, cuñete
kelp, laminaria (alga)
kelp forest, bosque de algas marinas
Kelvin, Kelvin
keratin, queratina
kernel, grano (semilla), núcleo (átomo)
kerosene, queroseno
ketone, cetona
kettle, tetera, caldera
kettle lake, lago de la hoya glaciar
keyboard (computer), teclado
kidney, riñón
kilo, kilo
kilobit, kilobit
kilobyte, kilobyte
kilogram (kg), kilogramo (kg)
kilohertz, kilohertz
kiloliter (kl), kilolitro (kl)
kilometer (km), kilómetro (km)
kilopascal, kilopascal
kilowatt, kilovatio
kilowatt hour, kilovatio hora
kinematics, cinemática
kinetic energy, energía cinética
kinetic molecular theory, teoría cinética molecular
kinetics, cinética
kingdom, reino
kitten, gatito
Klinefelter's syndrome, síndrome de Klinefelter
knapsack, mochila
knee, rodilla
knee joint, articulación de la rodilla
knee-jerk reflex, reflejo rotuliano
knob, perilla
knowledge, conocimiento
koala, koala
krill, krill
krypton, kriptón
Kuiper belt, cinturón de Kuiper

label[1], identificar, etiquetar

label[2], identificación, etiqueta
labor, parto
laboratory, laboratorio
laboratory burner, quemador de laboratorio
laboratory conclusion, conclusión de laboratorio
labyrinth, laberinto
laccolith, lacolito
lack, carecer de
lactase, lactasa
lactation, lactancia
Lactea, Láctea
lacteal, lácteo
lactic acid, ácido láctico
lactone, lactona
lactose, lactosa
ladybug, mariquita, catarina
lagoon, laguna
lake, lago
Lamarckism, Lamarckismo
lamella, lámina
lamina, tejido foliar
laminar flow, flujo laminar
lamprey, lamprea
lancelet, cefalocordado
lancet, lanceta
land, tierra
land breeze, brisa de tierra
land bridge, istmo
land mass, masa terrestre
land use, uso de la tierra
landfill, vertedero de basura
landform, accidente geográfico
landmass, masa continental
landscape, paisaje
landscape features, características del paisaje
landscape region, paisaje regional
landslide, avalancha de tierra, corrimiento de tierra
Langerhans, Langerhans
lanolin, lanolina
lanthanide, lantánido
lanthanide series, serie de los lantánidos
lanthanum, lantano
laptop, computadora portátil
large intestine, intestino grueso
larva, larva
laryngitis, laringitis
larynx, laringe
laser, láser
latent heat, calor latente
latent period, período de latencia
latent solvent, solvente latente
lateral erosion, erosión lateral
lateral fault, falla lateral
lateral line, línea lateral
lateral moraine, morrena lateral
latex, látex
latitude, latitud
latitudinal climate patterns, patrones latitudinales de clima
lattice, celosía
launch, lanzar, iniciar
lava, lava
lava flow, flujo de lava
lava plateau, meseta de lava
law, ley

law of action and reaction, ley de acción y reacción
Law of Constant Composition, Ley de la Composición Constante
law of dominance, ley de dominio
law of gravitation, ley de gravitación
law of independent assortment, ley de distribución independiente
Law of Reflection, Ley de Reflexión
law of segregation, ley de la segregación
law of use and disuse, ley del uso y desuso
lawn, césped
lawn mower, cortadora de césped
lawrencium, laurencio
laxative, laxante
layer, capa
layering, capas
LCD, pantalla de cristal líquido
leach, lixiviación
lead, plomo
lead chromate, cromato de plomo
leaf, hoja
leaf sheath, vaina de la hoja
leaflet, foliolo
leafstalk, peciolo
leak, filtrar
learn, aprender
learned, aprendido, erudito
learned adaptation, adaptación asimilada
learning, aprendizaje
learning disability, problema de aprendizaje
least, lo menor, lo mínimo
lecithin, lecitina
LED (light-emitting diode), LED (diodo emisor de luz)
ledge, reborde
leech, sanguijuela
leeward, sotavento
left-hand rule, regla de la mano izquierda
leg, pierna, pata, muslo
legend, leyenda
legume, legumbre
leguminous plant, planta leguminosa
lemur, lémur
length, longitud
length of a shadow, longitud de una sombra
lengthen, alargar
lens, lente
Lenz's Law, Ley de Lenz
leopard, leopardo
leprosy, lepra
lepton, leptón
leucine, leucina
leukemia, leucemia
leukocyte, leucocito
levee, dique
level, nivel
leveling forces, fuerzas de nivelación
lever, palanca
Leyden Jar, Leyden Jar
liberate, liberar
librarian, bibliotecario
library investigation, investigación bibliote-

caria
library references, referencias bibliotecarias
lichen, liquen
life, vida
life cycle, ciclo de vida
life expectancy, esperanza de vida
life-form, forma de vida
life science, ciencias de la vida
life span, duración de vida
lift[1]**,** levantar
lift[2]**,** fuerza de sustentación
ligament, ligamento
light, luz
light elements, elementos ligeros
light metal, metal ligero
light microscope, microscopio de luz
light wave, onda de luz
light-dependent reaction, reacción dependiente de la luz
light-emitting diode (LED), diodo emisor de luz (LED)
lighting, iluminación
lightning, relámpago
light-year, año luz
lignin, lignina
lignite, lignito
likewise, también
lily, lirio
lime, lima, limón
limestone, piedra caliza
limit, límite
limitations of information systems, limitaciones de los sistemas de información
limiting nutrient, limitación de nutrientes
limonite, limonita
line, línea
line graph, gráfico lineal
line of force, línea de fuerza
linear, lineal
linear accelerator, acelerador lineal
linear equation, ecuación lineal
linear momentum, momento lineal
liner, buque
link, enlace, eslabón
lip, labio
lipase, lipasa
lipid, lípido
lipid bilayer, bicapa lipídica
liquefaction, licuefacción
liquefy, licuar
liquid, líquido
liquid crystal, cristal líquido
liquid phase, fase líquida
liquor, licor
list[1]**,** lista
list[2]**,** enumerar
liter (L), litro (L)
literally, literalmente
literature review, revisión literaria
litharge, litargirio
lithium, litio
lithosphere, litósfera

lithospheric plates, placas litosféricas
litmus, tornasol
litmus paper, papel tornasol
little finger, dedo meñique
littoral, litoral
liver, hígado
liverwort, agrimonia
livestock, ganado
living, vivo
lizard, lagarto
load, carga
loam, limo
lobe, lóbulo
lobster, langosta
local energy needs, necesidades energéticas locales
local noon, mediodía local
location, lugar, ubicación
lock-and-key hypothesis, hipótesis de la cerradura-llave
lockjaw, trismo
locomotion, locomoción
lodestone, magnetita
loess, loess
log, tronco, leño
logarithm, logaritmo
logging, explotación forestal
logic, lógica
logistic growth, crecimiento logístico
longitude, longitud
longitudinal muscle, músculo longitudinal
longitudinal wave, onda de longitud, onda de longitudinal
longshore drift, deriva litoral
long-wave energy, energía en onda larga
loop of Henle, asa de Henle
loop samples, muestras de circuitos
loudness, ruidoso, estridente, estrepitoso, volumen
loudspeaker, altavoz
louse, piojo
low, bajo
low pressure (air front), presión baja
low temperature sink, sumidero de baja temperatura
low tide, marea baja
lowest common denominator, mínimo común denominador
lubricant, lubricante
lubricating oil, aceite lubricante
lumbar, lumbar
lumen, lumen
luminescent, luminiscente
luminous, luminoso
luminous flux, flujo luminoso
luminous intensity, intensidad luminosa
lunar, lunar
lunar eclipse, eclipse lunar
lunar module, módulo lunar
lung, pulmón
lungfish, pez pulmonado
luster, lustre, brillo
luteinizing hormone (LH), hormona luteinizante (HL)
lutetium, lutecio
lye, lejía

Lyme disease, enfermedad de Lyme
lymph, linfa
  lymph gland, ganglio linfático
  lymph node, ganglio linfático
  lymph vessel, vasos linfáticos
lymphatic system, sistema linfático
lymphocyte, linfocito
lyophilic, liofílico

machine, máquina
macroanalysis, macroanálisis
macroclimate, macroclima
macronucleus, macronúcleo
macula, mácula
mad cow disease, enfermedad de las vacas locas
mafic, máficas
magenta, magenta
magma, magma
magnalium, magnalium
magnesia, magnesia
magnesite, magnesita
magnesium, magnesio
magneson, magneton
magnet, imán
magnetic, magnético
  magnetic compass, brújula magnética
  magnetic declination, declinación magnética
  magnetic field, campo magnético
  magnetic field line, línea de campo magnético
  magnetic flux, flujo magnético
  magnetic force, fuerza magnética
  magnetic induction, inducción magnética
  magnetic pole, polo magnético
  magnetic resonance imaging (MRI), resonancia magnética nuclear (RMN)
  magnetic storm, tormenta magnética
magnetism, magnetismo
magnetite, magnetita
magnetize, imantar, magnetizar
magnetochemistry, magnetoquímica
magnetometer, magnetómetro
magnetron, magnetrón
magnification, exageración
magnifier, magnificador
magnify, agrandar, aumentar
magnifying glass, lupa
magnitude, magnitud
main sequence, secuencia principal
maintain, mantener, sostener
maintenance, mantenimiento
major, principal
malachite, malaquita
malaria, malaria
malathion, malatión
male, macho
malfunction, mal funcionamiento
malignant, maligno

malleability, maleabilidad
malnutrition, desnutrición
malocclusion, maloclusión
maltase, maltasa
maltose, maltosa
mammal, mamífero
mammary gland, glándula mamaria
mammogram, mamografía
mammoth, mamut
manage, dirigir, administrar, manejar
manatee, manatí
mandible, maxilar inferior, mandíbula
manganese, manganeso
manganese dioxide, dióxido de manganeso
manganic oxide, óxido de mangánico
mangrove, mangle, manglar
mangrove swamp, pantano de mangle
manipulated variable, variable manipulada
manner, modo
mannitol, manitol
manometer, manómetro
mantissa, mantisa
mantle, capa, manto
manufacture, fabricar
mar, estropear
marble, mármol
mare, yegua
margin, margen
marine, marino
marine biome, bioma marino
marine climate, clima marino
marine terrace, terraza marina
maritime polar airmass, masa de aire polar marítima
maritime tropical airmass, masa de aire tropical marítima
marrow, médula, tuétano
Mars, Marte
marsh, marisma
marsupial, marsupial
maser, máser
mass, masa
mass action, acción de masas
mass defect, defecto de masa
mass number, número de masa
mass production, fabricación en serie
mass spectograph, espectrógrafo de masas
massive, masivo
mastication, masticación
mastodon, mastodonte
mastoid, mastoideo
material, material
materials scientist, científico de materiales
maternal immunity, inmunidad materna
mathematical science, ciencias matemáticas
mathematically, matemáticamente
mathematician, matemático
mathematics, matemáticas
mating, apareamiento
matrix, matriz
matter, materia
matter wave, onda de materia
maturation, maduración
mature soil, suelo maduro

maturity, madurez
maximum, máximo
meadow, prado
mean, media (promedio)
  mean solar day, día solar medio
  mean time, tiempo medio
meander, deambular, serpentear
measles, sarampión
measurable, mensurable
measure[1], medir
measure[2], medida
measurement, medición
mechanical, mecánico
  mechanical advantage, ventaja mecánica
  mechanical energy, energía mecánica
  mechanical engineering, ingeniería mecánica
  mechanical force, fuerza mecánica
  mechanical weathering, desgaste mecánico
mechanics, mecánica
mechanism, mecanismo
media, medios de comunicación
medial moraine, morrena media
median, mediana
medical, médico
medicine, medicina
  medicine dropper, pipeta
medium, medio
medulla oblongata, bulbo raquídeo
medusa, medusa
mega, mega
megabit, megabit
megabyte, megabyte
megahertz, megahertz
megavolt, megavoltio
megawatt, megavatio
meiosis, meiosis
melanin, melanina
melanocyte cell, célula melanocita
melanoma, melanoma
melt, fundir
meltdown, fusión
melting, fusión
  melting point, punto de fusión
meltwater, agua de deshielo
member, miembro
membrane, membrana
membranous, membranoso
memory, memoria
menarche, menarquía
mendelevium, mendelevio
Mendelism, Mendelismo
meningitis, meningitis
meniscus, menisco
menopause, menopausia
menses, menstruación
menstrual cycle, ciclo menstrual
menstruation, menstruación
mental illness, enfermedad mental
menthol, mentol
mention, mencionar
mercury[1], mercurio
  mercury barometer, barómetro de mercurio
Mercury[2], Mercurio
meridian, meridiano
mesa, meseta
mesentery, mesenterio
mesh, engranar
mesoderm, mesodermo

meson, mesón
mesopause, mesopausa
mesophyll, mesófilo
mesophyte, mesófito
mesosphere, mesósfera
Mesozoic Era, Era Mesozoica
messenger RNA (mRNA), ARN mensajero
metabolic, metabólico
  metabolic waste, desechos metabólicos
metabolism, metabolismo
metacarpal, metacarpiano
metal, metal
  metal runner, riel de metal, guía de metal
metallic, metálico
  metallic bond, enlace metálico
  metallic conduction, conducción metálica
metalloid, metaloide
metallurgy, metalurgia
metamorphic, metamórfico
  metamorphic rock, roca metamórfica
metamorphism, metamorfismo
metamorphosis, metamorfosis
metaphase, metafase
metastasis, metástasis
metazoan, metazoario
meteor, meteoro
meteorite, meteorito
meteoroid, meteoroide
meteorology, meteorología
meter (m), metro (m)
  meter stick, varilla métrica
methane, metano
methanol, metanol
method, método
methyl, metilo
  methyl alcohol, alcohol metílico
methylamine, metilamina
methylene blue, azul de metileno
meticulously, meticulosamente
metric, métrico
  metric ruler, regla métrica
  metric system, sistema métrico
mica, mica
mice, ratones
micro-, micro-
microanalysis, microanálisis
microbe, microbio
microbiology, microbiología
microclimate, microclima
microcline, microclina
microdissection, microdisección
microfarad, microfaradios
microfilament, microfilamento
micrograph, micrografía
micrometer, micrómetro
  micrometer caliper, micrómetro
micronucleus, micronúcleo
microorganism, microorganismo
microphone, micrófono
microprocessor, microprocesador
microscope, microscopio
microscopic, microscópico
microsecond, microsegundo
microtubule, microtúbulo
microwave, microondas
midbrain, mesencéfalo, cerebro medio

middle ear, oído medio
middle finger, dedo del corazón
mid-latitude cyclone, ciclones de latitudes medias
mid-ocean ridge, dorsal oceánica
migraine, migraña
migration, migración
migratory, migratorio
mil, milésimo
mildew, moho
mile, milla
milk, leche
  milk tooth, diente de leche
Milky Way, Vía Láctea
milli, mili
milliampere, miliamperios
millibar, milibar
milligram (mg), miligramo (mg)
milliliter (ml), mililitro (ml)
millimeter (mm), milímetro (mm)
millipede, milpiés
milt, lecha seminal de los peces, líquido seminal de los peces
mimicry, mimetismo
mineral, mineral
  mineral acid, ácido mineral
  mineral oil, aceite mineral
  mineral water, agua mineral
minimum, mínimo
minute, minuto
mirage, espejismo
mirror, espejo
miscible, miscible
Mission Control, control de la misión
mite, ácaro
mitochondria, mitocondrios
mitochondrion (pl. mitochondria), mitocondrio
mitosis, mitosis
mitotic cell division, división celular mitótica
mix, mezclar, combinar
mixed, mezclado
  mixed number, número mixto
mixture, mezcla
moa, moa
mode, moda
model, modelo
modem, módem
moderator, moderador
modification, modificación
modified Mercalli scale, escala de Mercalli modificada
modify, modificar
modulate, modular
module, módulo
Moho, Moho
Mohorovicic discontinuity, discontinuidad de Mohorovicic
moist adiabatic lapse rate, tasa húmeda adiabática
moisture, humedad
molal boiling point constant, constante molal ebulloscópica
molal freezing point constant, constante molal crioscópica
molality, molalidad
molar, muela

mold, moho
mole, topo
molecular, molecular
    molecular formula, fórmula molecular
    molecular mass, masa molecular
    molecular sieve, tamiz molecular
    molecular weight, peso molecular
molecule, molécula
mollusca, moluscos
mollusk, molusco
molt, muda
molybdenum, molibdneo
momenta, momentos
momentum, momento, cantidad de movimiento
Monera, mónera
mongoose, mangosta
monitor, monitor
monkey, mono
monochromatic, monocromático
monoclinic, monoclínico
monocotyledon, monocotiledóneo, monocotiledóneas
monoculture, monocultivo
monocyte, monocitos
monodactyl, monodáctilo
monomer, monómero
monomial, monomio
monomineralic, monominerálico
monomolecular, monomolecular
monoploid, monoploide
monosaccharide, monosacárido
monosodium glutamate, glutamato monosódico
monounsaturated, monoinsaturado
monoxide, monóxido
monsoon, monzón
month, mes
moon, luna
    moon landing, alunizaje
moorings, amarras
moraine, morrena
morphine, morfina
morphology, morfología
mortality, mortalidad
morula, mórula
mosaic, mosaico
mosquito, mosquito
moss, musgo
moth, polilla
motherboard (computer), placa base
motile, motilidad
motility, motilidad
motion, movimiento
motor, motor
    motor action, acto motor
    motor nerve, nervios motores, centrífugos
    motor neuron, neurona motora
mountain, montaña
    mountain chain, sierra
    mountain lion, puma
    mountain peak, cumbre
    mountain range, cadena montañosa
mouse[1] (pl. mice), ratón
mouse[2] (computer), ratón, mouse

mouse pad, alfombrilla de ratón
mouth, boca
mouthpiece, boquilla
movable pulley, polea móvil
movement, movimiento
mower, cortacésped
MRI (magnetic resonance imaging), RMN (resonancia magnética nuclear)
mRNA (messenger RNA), ARNm (ARN mensajero)
mucous membrane, membrana mucosa
mucus, mucosa
mud, lodo
multicellular, multicelular
multiple[1], múltiplo
multiple[2], múltiple
multiple birth, parto múltiple
multiple choice question, pregunta de opción múltiple
multiple resistance, multirresistencia
multiple sclerosis, esclerosis múltiple
multiple-gene inheritance, herencia poligénica
multiplicand, multiplicando
multiplication, multiplicación
multiplier, multiplicador
multiply, multiplicar
mumps, paperas
muon, muón
muscle, músculo
muscle contraction, contracción muscular
muscle dystrophy, distrofia muscular
muscle fatigue, fatiga muscular
muscle tissue, tejido muscular
muscovite, moscovita
muscular, muscular
muscular system, sistema muscular
mushroom, hongo, seta
mussel, mejillón
mutagen, mutágeno
mutagenic agent, agente mutagénico
mutant, mutante
mutate, mutar
mutated, mutado
mutation, mutación
multicellular, multicelular
mutualism, mutualismo
mycelia, micelio
mycelium, micelio
mycology, micología
myelin, mielina
myelin sheath, vaina de mielina
myofibril, miofibrilla
myopia, miopía

N

nadir, nadir
nail[1], clavo
nail[2], uña
naked, desnudo
nano, nano
nanometer, nanómetro
nanosecond, nanosegundo
nanotube, nanotubo
naphtha, nafta
naphthalene, naftalina
narcotic, narcótico

narcotic drug, estupefaciente
nasal, nasal
nasal cavity, cavidad nasal
natural, natural
natural disaster, desastre natural
natural gas, gas natural
natural immunity, inmunidad natural
natural levees, diques naturales
natural logarithm, logaritmo natural
natural mother, madre biológica
natural number, número natural
natural resource, recurso natural
natural sciences, ciencias naturales
natural selection, selección natural
naturalist, naturalista
naturalize, naturalizar
naturally, naturalmente
nature, naturaleza
nature and nurture controversy, controversia entre natura y nurtura
nature of the surface, naturaleza de la superficie
nautical mile, milla náutica
nautilus, nautilo
neap tide, marea viva y muerta
nearby, cercano
nearsighted, miope
nearsightedness, miopía
nebula, nebulosa
neck, cuello
nectar, néctar
needle, aguja
negative, negativo
negative charge, carga negativa
negative feedback, retroalimentación negativa
nematocyst, nematocisto
nematode, nematodo
neodymium, neodimio
neon, neón
neoprene, neopreno
nephron, nefrona
Neptune, Neptuno
neptunium, neptunio
Nernst distribution law, ley de distribución de Nernst, ley de reparto
Nernst equation, ecuación de Nernst
nerve, nervio
nerve cell, neurona
nerve center, centro neurálgico
nerve cord, cordón nervioso
nerve fiber, fibra nerviosa
nerve impulse, impulso nervioso
nerve net, red nerviosa
nervous system, sistema nervioso
net force, fuerza resultante
net potential, red potencial
neural, neural
neural plate,

placa neural
neurohormone (neurotransmitter), neurohormona (neurotransmisor)
neurology, neurología
neuromuscular junction, unión neuromuscular
neuron, neurona
neurotransmitter, neurotransmisor
neutral, neutral
  neutral red, rojo neutro
  neutral solution, solución neutra
neutralization equivalent, neutralización equivalente
neutrino, neutrino
neutron, neutrón
  neutron star, estrella de neutrones
neve (firn), nevé (neviza)
new moon, luna nueva
newt, tritón, salamandra acuática
Newton, Newton
  Newton's First Law of Motion, Primera Ley de Movimiento de Newton
  Newton's Law of Gravitation, Ley de Gravitación de Newton
  Newton's Laws of Motion, Leyes de Movimiento de Newton
  Newton's Second Law of Motion, Segunda Ley de Movimiento de Newton
  Newton's Third Law of Motion, Tercera Ley de Movimiento de Newton
niacin, niacina
niche, nicho
nickel, níquel
nicotine, nicotina
night blindness, ceguera nocturna
nimbostratus, nimbostratus, nimboestrato
nimbus, nimbus
niobium, niobio
nipple, pezón
niter, nitro
nitrate, nitrato
nitration, nitración
nitric acid, ácido nítrico
nitric anhydride, anhídrido nitroso
nitric oxide, óxido nítrico
nitride, nitruro
nitriding, nitruración
nitrification, nitrificación
nitrifying bacteria, bacterias nitrificantes
nitrile, nitrilo
nitrite, nitrito
nitrogen, nitrógeno
  nitrogen family, familia del nitrógeno
  nitrogen-fixing bacteria, bacterias fijadoras de nitrógeno
nitrogenous, nitrogenado
  nitrogenous waste, residuos nitrogenados
nitroglycerin, nitroglicerina
nitrous acid, ácido nitroso
nitrous oxide, óxido nitroso
Nobel, Alfred B., Nobel, Alfred B.
nobelium, nobelio

noble gas, gas noble
nocturnal, nocturno
nodal, nodal
nodal line, línea nodal
node, nodo
nodule, nódulo
noise pollution, contaminación acústica
nomenclature, nomenclatura
noncommunicable, no transmisible
nonconformity, disconformidad
nondisjunction, disyunción
nonelectrolyte, no electrólito
nonliving thing, ser no vivo
nonmetal, no metálico
non-metallic, no metálico
nonperpendicular component, componente no perpendicular
nonperpendicular components of vector, componente no perpendicular de un vector
nonplacental mammal, mamífero sin placenta
nonpolar molecule, molécula no polar
nonrenewable, no renovable
nonrenewable energy resource, recurso energético no renovable
nonrenewable resource, recurso no renovable
nonsedimentary rock, roca no sedimentaria
nonvascular plant, plantas no vasculares
noradrenaline, noradrenalina, norepinefrina
normal boiling point, punto de ebullición normal
normal fault, falla normal
normal force, fuerza normal
normal salt, sal normal
normal solution, solución normal
normality, normalidad
North Pole, Polo Norte
North Star, Estrella Polar
Northern Hemisphere, hemisferio norte
northern lights, aurora boreal
nose, nariz
nostril, orificio nasal
notation, notación
note, advertir, observar
notochord, notocorda
nourish, nutrir, alimentar
nova, nova
nuclear, nuclear
nuclear bombardment, bombardeo nuclear
nuclear energy, energía nuclear
nuclear fission, fisión nuclear
nuclear fuel, combustible nuclear
nuclear fusion, fusión nuclear
nuclear magnetic resonance, resonancia magnética nuclear
nuclear membrane (envelope), membrana nuclear
nuclear model, modelo nuclear
nuclear physics, física nuclear
nuclear potential en-

ergy, energía potencial nuclear
nuclear power station, central nuclear
nuclear reaction, reacción nuclear
nuclear reactor, reactor nuclear
nuclear war, guerra nuclear
nuclear waste, vertidos nucleares
nuclear weapon, arma nuclear
nuclei, núcleos
nucleic acid, ácido nucleico
nucleoli, nucléolos
nucleolus, nucléolo
nucleon, nucleón
nucleotide, nucleótido
nucleus (pl. nuclei), núcleo
nuclide, nucleido
number, número
number lock (computer), bloqueo numérico
numeral, numeral
numerator, numerador
numerically, numéricamente
numerous, numerosos
nutrient, nutriente
nutrition, nutrición
nutritional, nutricional
nylon, nylon, nailon
nymph, ninfa

oak, roble
oasis, oasis
obey, obedecer
object, objetivo
objective, objetivo
objective lens, lente objetivo
oblate spheroid, esferoide oblato
oblique, oblicuo
oboe, oboe
observable, observable
observation, observación
observatory, observatorio
observe, observar
observed value, valor observado
obsidian, obsidiana
obstacle, obstáculo
obstetrics, obstetricia
obtain, obtener
obtained, obtenido
obtuse angle, ángulo obtuso
occasionally, ocasionalmente
occipital lobe, lóbulo occipital
occluded front, frente ocluido
occlusion, oclusión
occultation, ocultación
occupy, ocupar
occur, ocurrir
ocean, océano
ocean-floor spreading, expansión del fondo marino
oceanic crust, corteza oceánica
oceanica, oceánica
oceanography, oceanografía
octagon, octágono
octagonal, octagonal
octahedron, octaedro
octane, octano
octave, octavo
octet, octeto

octopus, pulpo
ocular, ocular
odd number, número impar
odor, olor
official, oficial
offset, offset, sistema de impresión
offspring, descendencia
ohm, ohmio
Ohm Law, Ley de Ohm
oil, aceite
old age, vejez
olefin, olefina
oleic acid, ácido oleico
olfaction, olfato
olfactory, olfativo, olfatorio
olfactory bulb, bulbo olfatorio, lóbulo olfativo
olfactory cell, célula olfativa
olfactory lobe, lóbulo olfativo
olfactory nerve, nervio olfativo
olivine, olivino
omnivore, omnívoro
oncogene, oncogén
oncology, oncología
one gene—one polypeptide hypothesis, hipótesis de un gen—un polipéptido
one-hole stopper, tapón de un orificio
ontogeny, ontogenia
onyx, ónice, ónix
oocyst, ooquistes
oocyte, ovocito
oogenesis, ovogénesis
ooze, fango, limo, cieno
opacity, opacidad
opal, ópalo
opaque, opaco
open, abierto
open chain, cadena abierta
open circuit, circuito abierto
open circulatory system, sistema circulatorio abierto
open cluster, cúmulo abierto
open field system, cultivo abierto
open universe, universo abierto
operating system, sistema operativo
operation, operación
operational definition, definición operativa
operculum, opérculo
operon, operón
ophthalmology, oftalmología
opinion, opinión
opium, opio
opossum, zarigüeya, tlacuache
opportunity, oportunidad
oppose, oponerse
optic, óptico
optic nerve, nervio óptico
optical, óptico
optical density, densidad óptica
optical fiber, fibra óptica
optical microscope, microscopio óptico
optical rotation, rotación óptica
optics, óptica

optimum, óptimo
oral, oral
orangutan, orangután
orbit[1], orbitar
orbit[2], órbita
orbital, orbital
  orbital pair, par de orbitales
  orbital speed, velocidad orbital
  orbital velocity, velocidad orbital
orca, orca
orchid, orquídea
order[1], orden
  order of reaction, orden de reacción
order[2], ordenar
ordinal number, número ordinal
ordinate, ordenada
Ordovician Period, Período Ordovícico
ore, mena
  ore deposit, depósito mineral
organ, órgano
  organ system, sistema de órganos
organelle, orgánulo
organic, orgánico
  organic acid, ácido orgánico
  organic chemistry, química orgánica
  organic compound, compuesto orgánico
organism, organismo
organize, organizar
organometallic, organometálica
orientation, orientación
origin, origen
original, original
  original horizontality, horizontalidad original
  original variables, variables originarias
Orion, Orión
ornithology, ornitología
orographic effect, efecto orográfico
orthoclase, ortoclasa
orthopedics, ortopedia
orthorhombic, ortorrómbico
oscillation, oscilación
oscilloscope, osciloscopio
osmium, osmio
osmosis, ósmosis
osmotic pressure, presión osmótica
ossification, osificación
osteoarthritis, osteoartritis
osteoblast, osteoblastos
osteocyte, osteocitos
osteology, osteología
osteoporosis, osteoporosis
ostrich, avestruz
otter, nutria
ounce, onza
outbreeding, exogamia
outcrop[1], aflorar
outcrop[2], afloramiento
outer core, núcleo externo
outer ear, oído externo
outlet, salida, desembocadura
output, salida
outward, exterior
outwash, acarreos fluvioglaciáricos
outwash plain, llanura fluvio glacial, llanura de aluvión
ova (sing. ovum), óvulos
oval, óvalo

ovaries, ovarios
ovary, ovario
overabundance, sobreabundancia, superabundancia
overall, en conjunto
overcast, cielo cubierto
overcome, vencer, superar
overfishing, sobrepesca
overgraze, sobrepastoreo
overload, sobrecarga
overlook, examinar
overpopulation, sobrepoblación
overtones, matices
oviduct, oviducto
oviparity, oviparidad
oviparous, ovíparo
ovipositor, ovipositor
ovoviviparous, ovovivíparo
ovulation, ovulación
ovule, óvulo
ovum (pl. ova), óvulo
owe, deber
owl, lechuza, búho
oxalic acid, ácido oxálico
oxbow, brazo muerto (de un río)
  oxbow lake, meandro
oxidation, oxidación
oxide, óxido
oxidize, oxidar
oxygen, oxígeno
  oxygen acid, oxígeno ácido
  oxygen consumption, consumo de oxígeno
  oxygen cycle, ciclo de oxígeno
  oxygen debt, deuda de oxígeno
  oxygen-carbon dioxide cycle, ciclo del oxigeno-dióxido de carbono
oxygenate, oxigenar
oxyhemoglobin, oxihemoglobina
oyster, ostra
oz., onza
ozone, ozono
  ozone layer, capa de ozono
  ozone shield, escudo de la capa de ozono

pacemaker, marcapasos
pachyderm, paquidermo
pack ice, hielo a la deriva
packing, embalaje
painstakingly, cuidadosamente
pair, par
paired, en pares
palate, paladar
Paleocene, Paleoceno
paleontology, paleontología
Paleozoic, Paleozoico
  Paleozoic Era, Era Paleozoica
palisade, empalizada
  palisade layer, capa en empalizada
  palisade mesophyll, mesófilo en empalizada
palladium, paladio
palm, palma
palmate, palmeado
pancreas, páncreas
pancreatic duct, conducto

pancreático
pancreatic juice, jugo pancreático
pandemic, pandemia
Pangaea, Pangea
panicle, panícula
papilla, papila
parabola, parábola
parabolic, parabólica
parachute, paracaídas
paraffin, parafina
paraffin series, serie de parafina
parallax, paralaje
parallel, paralelo
parallel circuit, circuito paralelo
parallel connection, conexión en paralelo
parallel force, fuerza paralela
parallel unconformity, discordancia paralela
parallelism, paralelismo
parallelogram, paralelogramo
paramecium, paramecio
parameter, parámetro
paraplegia, paraplejia
parasite, parásito
parasitic relationship, relación parasitaria
parasitism, parasitismo
parasympathetic nervous system (PNS), sistema nervioso parasimpático (SNP)
parathyroid gland, glándula paratiroides
parathyroid hormone, hormona paratiroidea
parenchyma, parénquima
parent cell, célula madre
parent generation, generación parental
parental care, cuidado parental
parental rock, roca parental
parietal lobe, lóbulo parietal
Parkinson's disease, enfermedad de Parkinson
parrot, loro
parsec, pársec
part, parte
parthenogenesis, partenogénesis
partial, parcial
partial pressure, presión parcial
partial product, producto parcial
particle, partícula
particle accelerator, acelerador de partículas
particular, particular
particulate, partícula
partition coefficient, coeficiente de reparto
parts per million, partes por millón
pascal, pascal, pascalio
Pascal Law, principio de Pascal
passerine, paseriforme
passive immunity, inmunidad pasiva
passivity, pasividad
pasteurization, pasteurización
patch, parche
patella, rótula
path, vía
pathogen, patógeno
pathogenic, patógenico, patógeno

pathology, patología
pattern, patrón
pavement, pavimento, escarpada
peacock blue, azul eléctrico
pearl, perla
pearly luster, brillo perlado
peat, turba
pebble, guijarro
pectin, pectina
pectoral, pectoral
peculiar, peculiar
peculiarity, peculiaridad
pediatrics, pediatría
pedicel, pedicelo
pedigree, raza, pedigrí, linaje
  pedigree chart, diagrama de pedigrí
peduncle, pedúnculo
peer review, revisión por pares
peer reviewed, revisados por pares, revisados por colegas
pegmatite, pegmatita
pelagic, pelágico
pelican, pelícano
pellagra, pelagra
pelvis, pelvis
pendulum, péndulo
peneplane, penillanura
penguin, pingüino
penicillin, penicilina
peninsula, península
penis, pene
pentagon, pentágono
pentane, pentano
penumbra, penumbra
pepsin, pepsina
peptic, péptico
  peptic ulcer, úlcera péptica
peptidase, peptidasa
peptide, péptido
  peptide bond, enlace peptídico
peptization, peptización
perceive, percibir
percent, por ciento, porcentaje
  percent error, porcentaje de error
percentage, por ciento, porcentaje
  percentage by mass, porcentaje de la masa
  percentage composition, porcentaje de la composición
percentile, percentil
perchlorate, perclorato
perennial, perenne
perfect, perfecto
  perfect number, número perfecto
perform, realizar
performer, ejecutante
perfume, perfume
pericardium, pericardio
peridotite, peridotita
perigee, perigeo
perihelion, perihelio
perimeter, perímetro
period, período
periodic, periódico
  periodic law, ley periódica
  periodic table, tabla periódica
peripheral nervous system, sistema nervioso periférico
periscope, periscopio
peristalsis, peristalsis
peritoneum, peritoneo
perjudicial disaster, de-

sastre perjudicial
permafrost, permafrost
permanent magnet, imán permanente
permanently, permanentemente
permeability, permeabilidad
permeable, permeable
Permian, Pérmico
  Permian Period, Período Pérmico
permit, permitir
peroxide, peróxido
perpendicular, perpendicular
  perpendicular force, fuerza perpendicular
perpendicularly, perpendicularmente
perspective, perspectiva
perspiration, transpiración
pest, peste, parásito
pesticide, pesticida
petal, pétalo
Petri dish, placa de Petri
petrifaction, petrificación
petrochemical, petroquímica
petroleum, petróleo
petrology, litología, petrología
pH, pH
  pH indicator, indicador de pH
  pH meter, pH-metro
phagocyte, fagocito
phagocytosis, fagocitosis
phalanges, falanges
pharmacology, farmacología
pharynx, faringe
phase, fase
  phase contrast microscope, microscopio de contraste de fase
  phase equilibrium, fase de equilibrio
phenol, fenol
phenolic, fenólicos
phenolphthalein, fenolftaleína
phenomena, fenómeno
phenomenon, fenómeno
phenotype, fenotipo
phenyl, fenil, fenilo
phenylalanine, fenilalanina
phenylketonuria (PKU), fenilcetonuria (PKU)
pheromone, feromona
philosopher, filósofo
phlegm, flema
phloem, floema
phlogiston, flogisto
phosphate, fosfato
phosphor, fósforo
phosphorescence, fosforescencia
phosphorus, fósforo
phosphorylation, fosforilación
photic zone, zona fótica
photo resistor, fotorresistencia
photochemical oxidant, oxidante fotoquímico
photochemistry, fotoquímica
photodegradable, fotodegradable
photoelectric, fotoeléctrico
  photoelectric colorimeter, colorímetro fotoeléctrico
  photoelectric effect, efecto fotoeléctrico
photoflash, luz relámpago
photography, fotografía
photolysis, fotólisis

photometry, fotometría
photomicrograph, fotomicrografía
photon, fotón
photoreceptor, fotorreceptor
photosphere, fotósfera
photosynthesis, fotosíntesis
phototropism, fototropismo
photovoltaic, fotovoltaica
 photovoltaic cell, celda fotovoltaica
phrenology, frenología
phycoerythrin, ficoeritrina
phylum, filum
physical, físico
 physical change, cambio físico
 physical chemistry, fisicoquímica
 physical environment, entorno físico
 physical equilibrium, equilibrio físico
 physical examination, reconocimiento médico
 physical model, modelo físico
 physical phenomena, fenómenos físicos
 physical property, propiedad física
 physical science, ciencias físicas
 physical weathering, erosión física, desgaste físico
physicist, físico
physics, física
physiology, fisiología
physiosorption, fisisorción
physiotherapy, fisioterapia
phytoplankton, fitoplancton

pi, pi
pico, pico
picofarad, picofaradios
pie chart, diagrama circular
pie graph, gráfica de pastel
pier, muelle
piezoelectric effect, efecto piezoeléctrico
piezoelectricity, piezoelectricidad
pig, cerdo
pigment, pigmento
pile, pila
pill, píldora
pimple, espinilla
pincers, tenazas
pine, pino
 pine cone, piña
 pine needle, aguja de pino
 pine nut, piñón
pineal, pineal
 pineal body, glándula pineal
 pineal gland, glándula pineal
pinna, pabellón auricular
pinnate, pinnado
pinocytic vesicle, vesícula pinocítica
pinocytosis, pinocitosis
pint, pinta
pioneer, pionero
 pioneer species, especie pionera
pipe, tubería
pipe still, alambique
pipette, pipeta
Pisces, Piscis
pistil, pistilo
pistol, pistola

piston, pistón
 piston release force, fuerza de liberación del pistón
 piston travel force, fuerza de desplazamiento del pistón
pit, pozo de mina
 pit viper, crotalino
pitch, tono
pith, médula
 pith ball, bolita
pituitary dwarfism, enanismo hipofisario
pituitary gland, glándula pituitaria
pixel, píxel
placebo, placebo
placenta, placenta
placental mammal, mamífero placentario
placental membranes, membranas de la placenta
placer, placel
plagioclase, plagioclasa
plague, plaga
plain, llanura
plane, plano
 plane geometry, geometría plana
 plane mirror, plano de simetría
 plane polarized, plano de polarización
planet, planeta
planetary nebula, nebulosa planetaria
planetary winds, vientos planetarios
plankton, plancton
plant, planta
Plantae, reino de las plantas
plaque, placa
plasma, plasma
 plasma cell, célula plasmática, célula plastocito
 plasma membrane, membrana plasmática
plasmid, plásmido
plasmodium, plasmodium
plastic, plástico
plasticizer, plastificante
plastid, plasto, plastidio
plate, plato, placas
 plate tectonics, tectónica de placas
 plate theory, teoría de las placas
plateau, meseta, altiplanicie
platelet, plaqueta
platform, plataforma
platinum, platino
platyhelminthes, platelmintos
platypus, ornitorrinco
Pleistocene, Pleistoceno
pleura, pleura
 pleural cavity, cavidad pleural
plexus, plexo
Pliocene, Plioceno
plot, parcela
plumage, plumaje
plume, pluma
plumule, plúmula
plunge, sumergirse
Pluto, Plutón
pluton, plutón
plutonium, plutonio
pluvial, pluvial
pneumatic, neumático
pneumonia, neumonía

poaching, caza furtiva
pod, vaina
pOH, potencial de oxidrilo
point, punto
point mutation, mutación puntual
poison, veneno, envenenar
poisonous, venenoso
polar, polar
    polar bond, enlace polar
    polar circle, círculo polar
    polar molecule, molécula polar
    polar zone, zona polar
Polaris, Polaris
polarity, polaridad
polarization, polarización
    polarization potential, potencial de polarización
polarize, polarizar
polarized, polarizado
polarizer, polarizador
polarographic, polarográfica
    polarographic analysis, análisis polarográfico
    polarographic apparatus, aparato polarográfico
    polarographic wave, onda polarográfica
pole, polo
polio, poliomielitis
pollen, polen
    pollen grain, grano de polen
    pollen tube, tubo de polen
pollinate, polinizar
pollination, polinización
pollutant, contaminante
pollution, contaminación, polución
polonium, polonio
poly-, poli-
polybasic acid, ácido polibásico
polycarbonate, policarbonato
polychloroprene, policloropreno
polychromatic, policromático
polycondensation, policondensación
polycyclic, policíclicos
polyelectrolyte, polielectrolito
polyester, poliéster
polyethylene, polietileno
    polyethylene glycols, polietilenglicoles
polygenic, poligénica
polyglycol, poliglicol
polygon, polígono
polyhedron, poliedro
polyhydric alcohol, polialcohol
polyisoprene, poliisopreno
polymer, polímero
polymerase chain reaction (PCR), reacción en cadena de la polimerasa
polymerization, polimerización
polymineralllic, polimineralica
polymorphism, polimorfismo
polynomial, polinomio
polyp, pólipo
polypetide, polipéptido

polyploidy, poliploidía
polypropylene, polipropileno
polysaccharide, polisacárido
polystyrene, poliestireno
polytetraflouro ethylene, politetrafluoroetileno
polyunsaturated, poliinsaturado
polyurethane, poliuretano
polyvinyl acetate, acetato de polivinilo
polyvinyl alcohol, alcohol de polivinilo
polyvinyl chloride, polivinílico
polyvinyl ether, éter de polivinilo
pond, estanque
population, población
population genetics, genética de poblaciones
porcelain, porcelana
pore, poro
porifera, poríferos
porometric, porométrico
porosity, porosidad
porous, poroso
porphyry, pórfido
porpoise, marsopa
port (computer), puerto (computadora)
portion, porción, parte
position, posición
positive, positivo
positive charge, carga positiva
positive feedback, retroalimentación positiva
positron, positrón, antielectrón
possess, poseer
posterior, posterior
postulate, postulado
pot, olla, maceta
potash, potasa
potassium, potasio
potassium hydroxide, hidróxido de potasio
potassium nitrate, nitrato de potasio
potassium permanganate, permanganato de potasio
potential, potencial
potential energy, energía potencial
potential evapotranspiration, evapotranspiración potencial
potentiometer, potenciómetro
potentiometric titration, valoración potenciométrica
pouched mammal, mamífero marsupial
pound, libra
pour, derramar
powder, polvo
powder metallurgy, metalurgia de polvos
power, potencia
practical, práctico
practically, prácticamente
prairie, pradera
praline, praliné
praseodymium, praseodimio
Precambrian Era, Era Precámbrica
precession, precesión
precipitant, precipitante
precipitate, precipitado

precipitation, precipitación
precipitation titration, titulación por precipitación
precise, preciso
precisely, precisamente
precision, precisión
predation, depredación
predator, predador, depredador
predator-prey relationship, relación depredador-presa
predatory, depredador
predict, predecir
prediction, predicción
predominate, predominar
prefix, prefijo
pregnancy, embarazo
prehensile, prensil
preliminary, preliminar
premature birth, parto prematuro
premolar, premolar
prepare, preparar
prepolymer, prepolímero
presence, presencia
present-day, hoy en día
preservative, conservante, preservativo
preserve, preservar
presoak, prelavado
pressure, presión
pressure gradient, gradiente de presión
presumably, presumiblemente
prevailing winds, vientos prevalecientes
prevent, prevenir
prevention, prevención
previous, previo
previously, previamente
prey, presa
prill, granulación
primarily, principalmente
primary, primario
primary alcohol, alcohol primario
primary coil, bobina primaria
primary color, color primario
primary consumer, consumidor primario
primary pigment, pigmento primario
primary productivity, productividad primaria
primary rock, roca primaria
primary root, raíz primaria
primary succession, sucesión primaria
primary wave, onda primaria
primate, primate
prime meridian, primer meridiano
prime number, número primo
primitive, primitivo
principal axis, eje principal
principal focal point, punto focal principal
principal focus, foco principal
principle, principio
principle of original horizontality, principio de la horizontalidad original
principle of superposition, principio de superposición

principle of uniformitarianism, principio del uniformismo
printer (computer), impresora
prism, prisma
probability, probabilidad
probability of occurrence, probabilidad de ocurrencia
probable, probable
probe, sonda
proboscis, probóscide
procedure, procedimiento
process, proceso
producer, productor
producer gas, gas pobre
product, producto
productivity, productividad
profile, perfil
progesterone, progesterona
program, programa
programmable, programable
progress made, avances realizados
prohibit, prohibir
project, proyecto
projectile, proyectil
projectile motion, movimiento de proyectiles
prokaryotic, procariota
promethium, prometeo
prominence, prominencia
prominent, prominente
promoter, promotor
proof, prueba
propane, propano
propel, impulsar
propeller, hélice
proper fraction, fracción propia
property, propiedad
prophase, profase
propionic acid, ácido propiónico
proportion, proporción
proportional, proporcional
proportionality, proporcionalidad
proportionality constant, proporcionalidad constante
proposal, propuesta
propose, proponer
propyl, propilo
propylene, propileno
prostaglandin, prostaglandina
prostate, próstata
prostate gland, glándula prostática
protactinium, protactinio
protease, proteasa, peptidasa
protect, proteger
protein, proteína
protest, protesta
prothrombin, protrombina
protist, protista
Protista, reino Protista
protocol, protocolo
protolysis reaction, reacción de protólisis
proton, protón
proton acceptor, aceptor de protones
proton donor, donante de protones
protoplasm, protoplasma
protozoan (pl. protozoa), protozoo (pl. protozoos), protozoario
protractor, transportador
prove, probar
provide, proveer

proximity, proximidad
Prussian Blue, azul de Prusia
pseudo pod, seudópodo
psychiatry, psiquiatría
psychology, psicología
psychosis, psicosis
psychrometer, psicrómetro
pterodactyl, pterodactiloideo
ptyalin, ptialina, amilasa
puberty, pubertad
pubis, pubis
public domain software, software de dominio público
puck, disco
puddle, charco
pulley, polea
pull-tab, lengüeta de tiro
pulmonary, pulmonar
  pulmonary artery, arteria pulmonar
  pulmonary circulation, circulación pulmonar
  pulmonary vein, vena pulmonar
pulp, pulpa
pulsar, púlsar
pulse, pulso
pumice, piedra pómez
pump, bomba
punctuated equilibrium, equilibrio interrumpido
Punnett square, diagrama de Punnett
pupa, crisálida
pupil (eye), pupila
puppy, cachorro
pure, puro
  pure dominant, puro dominante
  pure recessive, puro recesivo
purine, purina
purity, pureza
purple, púrpura
pus, pus
putty, masilla
P-wave, onda P
pyloric sphincter, esfínter pilórico
pyramid (math), pirámide
  pyramid of energy, pirámide de la energía
pyridoxine, piridoxina
pyrite, pirita
pyrolysis, pirólisis
pyroxene, piroxeno
pyrrole, pirrol
pyruvic acid, ácido pirúvico
Pythagorean, pitagórico
  Pythagorean theorem, teorema de Pitágoras
python, pitón, pitónido

quadrangular, cuadrangular
quadrant, cuadrante
quadratic, cuadrático
quadriceps, cuádriceps
quadrilateral, cuadrilátero
quadruped, cuadrúpedo
qualitative, cualitativo
  qualitative analysis, análisis cualitativo
quality, calidad
  quality control, control de calidad
quanta, cuantos de energía
quantitative, cuantitativo
  quantitative analysis, análisis cuantitativo

quantity, cantidad
quantize, cuantización
quantum, cuanto
  quantum mechanics, mecánica cuántica
  quantum number, número cuántico
  quantum theory, teoría cuántica
quarantine, cuarentena
quark, quark
  quark model nucleon, modelo quark de los nucleones
quart, cuarto (unidad)
quarter moon, cuarto de luna
quartz, cuarzo
quartzite, cuarcita
quasar, quásar
Quaternary Period, Período Cuaternario
queen, reina
  queen bee, abeja reina
query, consulta
question, pregunta
quick, rápido
quicksand, arenas movedizas
quill, pluma
quinine, quinina
quinone, quinona
quotation, cotización
quotient, cociente

R&D (research and development), I+D (investigación y desarrollo)
rabbit, conejo
rabies, rabia
raceme, racimo
racetrack, canal de conducción, pista
raceway, canal de conducción, pista
racquet, raqueta
radar, radar
  radar gun, pistola láser
radial, radial
  radial pattern, patrón radial
  radial symmetry, simetría radial
radially, radialmente
radian, radián
radiant, radiante
  radiant energy, energía radiante
radiating canal, canal radial
radiation, radiación
radiative balance, balance radiativo
radiator, radiador
radical, radical
radicand, radicando
radicle, radícula
radio, radio
  radio frequency, radiofrecuencia
  radio telescope, radiotelescopios
  radio wave, onda de radio
radioactive, radiactivo
  radioactive decay, desintegración radiactiva
  radioactive element, elemento radiactivo
  radioactive material, material radiactivo
  radioactive series, serie radiactiva

radioactivity, radiactividad
radiocarbon, radiocarbono
radiocarbon method, método de datación por radiocarbono
radioisotope, radioisótopo
radiology, radiología
radiometer, radiómetro
radiotherapy, radioterapia
radium, radio (elemento químico)
radius[1], radio (geometría)
radius[2], radio
radon, radón
ragweed, ambrosía
rain, lluvia
rain forest, selva tropical
rain shadow, sombra de lluvia
rainbow, arco iris
rainbow trout, trucha arco iris
raise, aumentar
ramp, rampa, desnivel
random, aleatorio
range, rango, intervalo, alcance
rank, rango, grado
Rapid Eye Movement (REM), movimiento ocular rápido (MOR)
rapidly, rápidamente
raptor, ave rapaz
rare, excepcional, raro
rare earth, tierras raras
rare gas, gas raro
rare metal, metal raro
rarefaction, rarefacción
rate, índice, velocidad, razón
at the rate of, a razón de
rate of acceleration, tasa de aceleración
rate-determining step, paso determinante de la velocidad
ratio, relación
ratio sum, suma de cocientes
rational number, número racional
ray, rayo (física), semirrecta (matemáticas)
ray optics, óptica de rayos
razor, navaja, rasuradora
reabsorption, reabsorción
react, reaccionar
reactant, reactivo
reaction, reacción
reaction mechanism, mecanismo de reacción
reaction rate, velocidad de reacción
reactor, reactor
read-only memory (ROM), memoria de solo lectura (ROM)
readily, fácilmente
reagent, reactivo
real, real, verdadero
real image, imagen real
real number, número real
realistic, realista
rearrange, reordenar
rearrangement, reordenamiento, reorganización
reason, razón
rebound, rebote
recede, alejarse

receive, recibir
receiver, receptor
recently, recientemente
receptacle, receptáculo
receptor, receptor
receptor molecule, molécula receptora
recessional moraine, morrena de recesión
recessive, recesivo
recessive gene, gen recesivo
recharge, recarga
reciprocal, recíproco
reclaim, reclamar
recognize, reconocer
recombination, recombinación
recombination gamete, recombinación de gametos
recombine, recombinar
record, registrar, grabar
recover, recuperar
recrystallization, recristalización
rectangle, rectángulo
rectangular, rectangular
rectangular pattern, patrón rectangular
rectification, rectificación
rectilinear, rectilíneo
rectum, recto
recycle, reciclar
recycling, reciclaje
red, rojo
red blood cell, eritrocito
red corpuscle (red blood cell), glóbulo rojo (hematíes)
red dwarf, enana roja
red giant, gigante roja
red tide, marea roja
red wine, vino tinto
redefine, redefinir
redox reaction, reacción redox
redshift, desplazamiento hacia el rojo
reduce, reducir
reduced, reducido
reducing flame, reducción de la llama
reduction, reducción
reduction division (meiosis), regulado
redwood, madera roja
reed, caña
reef, arrecife
reestablish, restablecer
refer, referir
reference, referencia
reference point, punto de referencia
refinery, refinería
refining, refinación
refining zone, zona de refinación
reflect, reflejar, reflexionar
reflection, reflejo, reflexión
reflection telescope, telescopio reflector
reflex, reflejo
reflex arc, arco reflejo
reflex center, centro de reflejo
reflux, reflujo
reforestation, reforestación
refract, refractar

refracting telescope, telescopio refractor
refraction, refracción
refractory period, período refractario
refrigerant, refrigerante
refrigerator, refrigerador
refute, refutar
regard, considerar
regeneration, regeneración
region, región
register, registro
regolith, regolito
regular, regular
regular reflection, reflexión regular
regulate, regular
regulated, regularizado
regulation, regulación
regurgitation, regurgitación
reinforce, reforzar
reinforcement, refuerzo
reinforcing agent, agente de refuerzo
relate, relacionar
related, relacionado
relationship, relación
relative dating, datación relativa
relative humidity, humedad relativa
relatively, relativamente
relativity, relatividad
relay, retransmitir, relé
release, soltar, liberar, dar curso
relevant, relevante
reliable, confiable
rely on, confiar en
REM (Rapid Eye Movement), MOR (movimiento ocular rápido)
remain, permanecer
remain the same, permanece sin cambios
remainder, residuo (matemáticas)
remains, restos, sobras
remission, remisión
remora, rémora
remote, remoto, lejano
remove, quitar
renal, renal
renal artery, arteria renal
renal circulation, circulación renal
renal portal vein, vena porta renal
renal vein, vena renal
renewable, renovable
renewable energy, recurso energético renovable
renewable resource, recurso renovable
rennin, renina
repeat, repetir
repellent, repelente
repetitious, repetitiva
replace, reemplazar
replacement, reemplazo
replicate, replicar
reported cases, casos declarados
represent, representar
reproduce, reproducir
reproduction, reproducción
reproductive cell, gameto
reproductive isolation, aislamiento reproductivo
reproductive system, aparato reproductor
reptile, reptil
repulsion, repulsión

repulsive, repulsivo
required, requerido
research, investigación
  research plan, plan de investigación
researcher, investigador
resemble, parecerse
reservoir, embalse
residual sediment, sedimento residual
residual soil, suelo residual
residue, residuos
resin, resina
resist, resistir
resistance, resistencia
resistant strain, cepa resistente
resistant to, resistente a
resistor, resistor
resolution power of lens, poder de resolución de una lente
resolve, resolver
resonance, resonancia
resource, recurso
  resources management, gestión de los recursos
respective, respectivo
respiration, respiración
respiratory chain, cadena respiratoria
respiratory surface, superficie de las vías respiratorias
respiratory tract, vías respiratorias
respond, responder
response, respuesta
responsible, responsable
restatement, reafirmación
resting potential, potencial de reposo
resting stage (interphase), fase de reposo (interfase)
restore, restablecer, restaurar
restoring force, elasticidad
result[1], resultado
result[2], resultar
resultant, resultante
retain, conservar
reticulate, reticular
reticulum, retícula
retina, retina
retort, retorta
retrograde, retrógrado
  retrograde motion, movimiento retrógrado
retrovirus, retrovirus
reuse, reutilizar
reveal, revelar
reverse, reverso
  reverse fault, falla inversa
reversed polarity, polaridad invertida
reversible reaction, reacción reversible
review, revisión
revolution, revolución
revolve, girar, dar vueltas
Rh factor, factor Rh
Rh negative blood, sangre Rh negativo
Rh positive blood, sangre Rh positivo
rhenium, renio
rheostat, reóstato
Rhesus factor (Rh factor), factor Rhesus (factor Rh)
rheumatic fever, fiebre reumática
rheumatoid arthritis, artritis reumatoide
rhinoceros, rinoceronte

rhizoid, rizoide
rhizome, rizoma
rhodium, rodio
rhombus, rombo
rhyolite, riolita
rib, costilla
rib cage, caja torácica
ribbon, cinta
riboflavin, riboflavina
ribonucleic acid (RNA), ácido ribonucleico (ARN)
ribose, ribosa
ribosomal RNA, ARN ribosomal
ribosome, ribosoma
Richter scale, escala de Richter
rickets, raquitismo
ridge, cresta
rift, falla
Rift Valley, valle del Rift
rift zone, zona de fractura, zona rift
right angle, ángulo recto
right ascension, ascensión recta
right triangle, triángulo rectángulo
rigid, rígido
rigor mortis, rigor mortis
ring, anillo
ring finger, dedo anular
ring stand, anillo de soporte
ring structure, estructura de anillo
ringing, zumbido
ringworm, tiña
rip, desgarrar, rasgar
rip current, corriente de resaca
ripple, onda
ripple tank, cubeta de ondas
rise, subir
risk, riesgo
river, río
rivet, remache
RNA, ARN
roach, cucaracha, escarabajo
robot, robot
robotics, robótica
rock, roca
rock cycle, ciclo de las rocas
rock flour, harina de roca
rock formation, formación rocosa
rock resistance, resistencia de la roca
rock salt, sal de piedra
rock-forming minerals, minerales formadores de roca
rocket, cohete
rocket engine, motor cohete
rocket launcher, lanzacohetes
rocking, balanceo
Rocky Mountain spotted fever, fiebre manchada de las Montañas Rocosas
rod, bastón (célula), rod (unidad de longitud)
rodent, roedor
roe, hueva
role, papel, rol
roll, rollo

rolling, rodante
ROM (read-only memory), ROM (memoria de solo lectura)
rookery, colonia de grajos
root, raíz
  root hair, raíz capilar
  root nodule, nódulos radicales
  root pressure, presión de raíz
rope, cuerda
rotate, rotar
rotation, rotación
  rotation of crops, rotación de cultivos
rotator cuff, manguito rotador
rotifer, rotífero
rough, tosco, desigual
roughage, forraje
roundworm, ascáride
router, router, enrutador, direccionador
routine, rutina
rubber, hule
  rubber ball, pelota de goma
rubbing, fricción
rubella, rubéola
rubidium, rubidio
ruby, rubí
rudimentary organ, órganos rudimentarios
ruler, regla graduada
ruminant, rumiante
runner, corredor
runoff, residuo líquido
rush, apuro, urgencia
rust, óxido, herrumbre
ruthenium, rutenio

Sabin vaccine, vacuna Sabin
sac, saco, bolsa membranosa
saccharide, sacárido
saccharin, sacarina
saccharose, sacarosa
sack, saco
sacrum, sacro
sag, hundimiento
Sagittarius, Sagitario
sailor, marinero
sal soda, sal de soda
salamander, salamandra
salicylic acid, ácido salicílico
saline, salino
salinity, salinidad
saliva, saliva
salivary amylase, amilasa salival
salivary gland, glándula salival
salivate, salivar
Salk vaccine, vacuna Salk, vacuna de poliovirus inactivados
salmon, salmón
salmonella, salmonella
salt, sal
  salt bridge, puente salino
  salt marsh, salina
  salt mine, salina
  salt spring, fuente de agua salada
saltation, saltación
saltwater, agua salada
samarium, samario
sample, muestra

sand, arena
sandbar, banco de arena
sandstone, arenisca
sandy soil, suelo arenoso
sanitary, sanitario
    sanitary landfills, rellenos sanitarios
sap, savia
saponification, saponificación
sapphire, zafiro
saprophyte, saprófito
saprophytism, saprofitismo
sarcoma, sarcoma
satellite, satélite
saturated, saturado
    saturated fat, grasa saturada
    saturated solution, solución saturada
saturation, saturación
    saturation point, punto de saturación
    saturation vapor pressure, presión de vapor a saturación
    saturation zone, zona de saturación
Saturn, Saturno
savanna, sabana
saxophone, saxófono
scaffold, andamio
scalar, escalar
    scalar field, campo escalar
scale, escala, balanza
    scale insect, cochinilla
scalene, escaleno
scallop, venera
scandium, escandio
scapula, escápula
scarab, escarabajo
scarce, escaso
scarcity, escasez
scarlet fever, escarlatina
scattering, dispersión
scavenger, carroñero
schematic, esquemática
schist, esquisto
schizophrenia, esquizofrenia
Schwann's cell, células de Schwann
sciatic nerve, nervio ciático
science, ciencia
scientific, científico
    scientific data, datos científicos
    scientific inquiry, indagación científica
    scientific investigation, investigación científica
    scientific law, ley científica
    scientific literacy, alfabetización científica
    scientific model, modelo científico
    scientific name, nombre científico
    scientific notation, notación científica
    scientific theory, teoría científica
    scientific thinking, razonamiento científico
scientist, científico
scintillation, centelleo
scion, vástago
scissors, tijeras
sclera, esclerótica
sclereid, esclereidas
scoliosis, escoliosis
scoria, escoria
scorpion, escorpión

scrape[1], raspar, rascar
scrape[2], raspadura
scratch, rayar, retirar
screen, proyectar
screw[1], atornillar
screw[2], tornillo
screwdriver, destornillador
scroll (computer), desplazar el cursor
  scroll down, bajar el cursor
  scroll up, subir el cursor
scrotum, escroto
scuba, escafandra autónoma
scurvy, escorbuto
sea, mar
  sea anemone, anémona de mar
  sea bass, lubina
  sea breezes, brisa marina
  sea cliff, acantilado
  sea cucumber, cohombro de mar
  sea floor spreading, expansión del fondo marino
  sea gull, gaviota
  sea horse, caballo marino
  sea level, nivel del mar
  sea lion, león marino
  sea urchin, erizo de mar, equino
seal, foca
sealant, sellador
seam, veta, filón, grieta
seamount, monte submarino
search, búsqueda
season, estación
seasonally, estacionalmente
seaweed, alga marina
sebaceous gland, glándula sebácea
sebum, sebo
secant, secante
second, segundo(a)
  second filial generation, segunda generación filial
  Second Law of Thermodynamics, Segunda Ley de la Termodinámica
  second left-hand rule, segunda regla de la mano izquierda
  second-order line, línea de segundo orden
secondary, secundario
  secondary alcohol, alcohol secundario
  secondary coil, bobina secundaria
  secondary color, color secundario
  secondary consumer, consumidor secundario
  secondary mycelium, micelio secundario
  secondary pigment, pigmento secundario
  secondary sexual characteristics, características sexuales secundarias
  secondary succession, sucesión secundaria
  secondary wave, onda secundaria

secrete, secretar
secretin, secretina
secretion, secreción
sector, sector
sedative, sedante, sedativo
sediment, sedimento
sedimentary, sedimentario
 sedimentary rock, roca sedimentaria
 sedimentary strata, estratos sedimentarios
sedimentation, sedimentación
 sedimentation balance, equilibrio de sedimentación
 sedimentation potential, sedimentación potencial
see, ver, entender
seed, semilla
 seed crystal, cristal semilla
 seed dispersal, dispersión de semillas
seedling, cultivo en semillero
seep, filtrar, escurrir
segment, segmento
segregation, segregación
seismic, sísmico
 seismic exploration, exploración sísmica
 seismic wave, onda sísmica
seismograph, sismógrafo
seismology, sismología
select[1], seleccionar
select[2], selecto
selection, selección
selective breeding, reproducción selectiva
selective permeability, permeabilidad selectiva
selectively permeable membrane, membrana permeable selectiva
selenium, selenio
self-fertilization, autofecundación
self-inductance, autoinductancia
self-pollination, autopolinización
semen, semen
semiarid, semiárido
semicircle, semicírculo
semicircular canal, canal semicircular
semiconductor, semiconductor
semilunar valve, válvula semilunar
semimicroanalysis, semimicroanálisis
seminal vesicle, vesícula seminal
seminiferous tubule, túbulos seminíferos
semipermeable, semipermeable
semipermeable membrane, membrana semipermeable
semisynthetic, semisintético
sensation, sensación
sense[1], sentir
sense[2], sentido
 sense organ, órgano del sentido
sensitive, sensible
sensitivity, sensibilidad
sensitization, sensibilización
sensor, sensor
sensory, sensorial
 sensory area, área

sensorial
sensory nerve fiber, fibras nerviosas sensoriales
sensory neuron, neurona sensorial
sensory organs, órganos sensoriales
sensory receptor, receptor sensorial
sepal, sépalo
separate, separar
separated, separado
separation, separación
sepsis, sepsis
septum, tabique nasal
sequence, sucesión, secuencia
sequestering agent, agente de secuestro
series, serie
series circuit, circuito en serie
series connection, conexión en serie
series-parallel circuit, circuito en serie paralelo
serum, suero
served as, servir como
server, servidor
sessile, sésil
set[1], fijo
set[2], conjunto
set[3], grupo círculo
set[4], juego
set[5], encargar
set[6], fijar
settle, asentar, posar, depositar
setup, instalación
sewage, aguas residuales
sewage treatment, tratamiento de aguas residuales
sewing, coser
sewing needle, aguja de coser
sex, sexo
sex cells, gametos
sex change, cambio de sexo
sex chromosome, cromosoma sexual
sex education, educación sexual
sex hormone, hormona sexual
sex organ, órgano sexual
sex-linkage inheritance, herencia ligada al sexo
sex-linked gene, gen vinculado al sexo
sexual, sexual
sexual discrimination, discriminación sexual
sexual generation, generación sexual
sexual harassment, acoso sexual
sexual intercourse, el acto sexual, coito
sexual maturity, madurez sexual
sexual reproduction, reproducción sexual
sexually transmitted disease (STD), enfermedad de transmisión sexual (ETS)
shale, esquisto
shallow, superficial
shape[1], forma

shape[2]**,** formar
share, compartir
shark, tiburón
shelf, estante, plataforma
shell, cáscara, capa, concha
shield, escudo
  shield cone, escudo de cono
  shield volcano, volcán en escudo
shift key (computer), tecla de mayúsculas
ship builder, constructor de barcos
shirt, camisa
shiver, tiritar, temblar
shoal, banco de arena
shock, choque, sacudida, conmoción
  shock absorber, amortiguador
  shock therapy, terapia de choque, terapia de electrochoque
shockwave, onda expansiva
shoot, retoño, vástago
shooting star, estrella fugaz
shore, orilla
shoreline, orilla
short circuit, cortocircuito
shortsightedness, miopía
shot-putter, lanzador de peso
shoulder, hombro, acotamiento, banquina
  shoulder joint, articulación del hombro
shovel, pala
shrink, encoger, contraer
shrub, arbusto
  shrub layer, capa de arbusto
Siamese twins, gemelos siameses
sibling, hermano
sickle cell, célula falciforme
  sickle cell anemia, anemia falciforme
side effect, efecto secundario
sideral day, día sideral
sideral month, mes sideral
sidereal, sidéreo
sidewalk, acera
sideway, lateral
sierra, sierra
significance, importancia
significant, significativo
  significant digits, dígitos significativos
silica, sílice
  silica gel, gel de sílice
silicate, silicato
silicon, silicón, silicio
  silicon carbide, carburo de silicio
  silicon chip, chip de silicio
silicone, silicona
  silicone oil, aceite de silicona
silicon-oxygen tetrahedron, tetraedro de silicio-oxígeno
silk, seda
silkworm, gusano de la seda
sill, alféizar
silt, cieno
siltstone, limolita
Silurian, Siluriano
silver, plata
  silver alloy, aleación de plata
  silver nitrate, nitrato de plata
  silver plate,

baño de plata
silver-thistle, acanto, branca ursina
silver-weed, agrimonia
similarity, semejanza
simple, simple
simple fraction, fracción simple
simple harmonic motion, movimiento armónico simple
simple machine, máquina simple
simple microscope, microscopio simple
simple reflex action, acción de reflejo simple
simple sugar (monosaccharide), azúcares simples (monosacáridos)
simplify, simplificar
sine, seno
single circulation, circulación simple
single-gene trait, gen recesivo
single-slit diffraction, difracción de una sola rendija
sink (energy), sumidero (energía)
sinkable, hundibles
sino-atrial node (S-A node), nódulo sinoauricular
sinter, sinterizar
sintered crucible, crisol de sinterizado
sinus, seno
sinusoidal projection, proyección sinusoidal
siphon, sifón
siren, sirena
Sirius, Sirio
site, sitio
situation, situación
size, tamaño
skating, patinaje
skeletal muscle, músculo esquelético
skeletal system, esqueleto
skeleton, esqueleto
skeleton key, llave maestra
sketch, bosquejo, croquis
skin, piel
skin cancer, cáncer de piel
skin graft, injerto de piel
skull, cráneo
sky, cielo
skydiver, paracaidista
slate, pizarra
sled, trineo
sleep[1], sueño
sleep[2], dormir
sleepwalker, sonámbulo
sleepwalking, sonambulismo
sleeping sickness, encefalitis letárgica
sleet, aguanieve
slide, deslizar
slide rule, regla de cálculo
slide value, válvula de corredera
slightly, ligeramente
slime, lodo mineral
slime mold, moho mucilaginoso
slit, cortar
slope, pendiente
slot (computer), ranura
slough, pantano, canal de agua
slow oxidation, oxidación lenta

sludge, lodo
slurry, estiércol líquido
small intestine, intestino delgado
smallpox, viruela
smash, estrellar
smell[1], olor, olfato
smell[2], oler
smelt[1], esperlán, eperlano
smelt[2], fundir
smog, niebla tóxica, esmog
smoke[1], humo
smoke[2], fumar, ahumar
smokestack, chimenea
smooth, suave
  smooth muscle, músculo liso, visceral
smut, tizón, hollín
snake, serpiente, culebra
snap, romper, chasquear
sneezing reflex, acto reflejo del estornudo.
Snell's Law, Ley de Snell
snout, hocico, morro
snow[1], nieve
  snow blindness, deslumbramiento por la nieve
  snow line, límite de las nieves perpetuas, línea de nieve
snow[2], nevar
snowmobile, moto de nieve
snowstorm, nevasca, nevada, nevisca
soak, empapar
soap, jabón
social science, ciencias sociales
sociology, sociología
socket, enchufe
soda, soda
  soda ash, ceniza de soda, carbonato de sodio
sodium, sodio
  sodium bicarbonate, bicarbonato sódico
  sodium carbonate, carbonato sódico
  sodium chloride, cloruro de sodio
  sodium fluoride, fluoruro de sodio
  sodium hydroxide, hidróxido de sodio
  sodium ion, iones de sodio
  sodium nitrate, nitrato de sodio
software, software
  software engineer, ingeniero de programas
  software package, paquete de software
soil, suelo
  soil association, asociación de suelos
  soil conservation, conservación del suelo
  soil depletion, agotamiento del suelo
  soil erosion, erosión del suelo
  soil horizon, horizontes del suelo
  soil profile, perfil del suelo
  soil solution, solución del suelo
  soil storage, almacenamiento del suelo
  soil texture, tipo de suelo, textura de suelo
sol, sol
solar, solar

solar cell, celda solar
solar eclipse, eclipse de sol
solar energy, energía solar
solar farm, huerta solar
solar flare, erupción solar
solar noon, mediodía solar
solar plexus, plexo solar
solar power, energía solar
solar system, sistema solar
solar time, tiempo solar
solar wind, viento solar
solar year, año solar
soldier ant, hormiga obrera
solenoid, solenoide
solid, sólido
solid bone, hueso sólido
solid phase, fase sólida
solid solution, solución sólida
solid state, estado sólido
solid waste, residuos sólidos
solidification, solidificación
solidify, solidificar
solidus, Sólidus
soliquoid, soliquoide
solstice, solsticio
solubility, solubilidad
solubility curve, curva de solubilidad
solubility product constant, constante del producto de solubilidad
solubility product expression, expresión del producto de solubilidad
soluble, soluble
solute, soluto
solution, solución
solution equilibrium, solución de equilibrio
solvent, solvente, disolvente
somatic, corporal, corpóreo, físico
somatic cell, células somáticas
somatic nervous system, sistema nervioso somático
sonar, sónar
sonic, sónico
sonic barrier, barrera sónica
sonic boom, estampido sónico
soot, hollín
soprano, soprano
sorption, sorción
sort, tipo
sorting, clasificación
sound, sonido
sound barrier, barrera del sonido
sound recorder, grabador de sonido
sound wave, onda de sonido
sour, agrio
source, fuente

source region, región geografía
South Pole, Polo Sur
Southern Cross, Cruz del Sur
Southern Hemisphere, hemisferio sur
space, espacio
space bar (computer), barra espaciadora
spacecraft, nave espacial
span, abarcar
spark, chispa
spawn, hueva
specialize, especializar
speciation, especiación
species, especie
specific, específico
specific gravity, gravedad específica
specific heat, calor específico
specificity, especificidad
specimen, especimen
speck, mota, partícula
spectacular, espectacular
spectra, espectros
spectral lines, líneas espectrales
spectrograph, espectrógrafo
spectrometer, espectrómetro
spectroscope, espectroscopio
spectroscopy, espectroscopia
spectrum, espectro
speed, velocidad
sperm, esperma
sperm duct, espermatozoides del conducto deferente
sperm nuclei, núcleos de los espermatozoides
spermaceti, espermaceti, esperma de ballena
spermatid, espermátide
spermatocyte, espermatocito
spermatogenesis, espermatogénesis
spermatophyte, espermatofita
spermatozoon, espermatozoide, espermátulo, espermatozoario
sphagnum, esfagno
sphalerite, esfalerita
sphere, esfera
spherical, esférica
spherical aberration, aberración esférica
spheroid, esferoide
sphincter, esfínter
spiderweb, telaraña
spike, espliego
spill[1], derrame
spill[2], derramar
spin, girar
spinal, espinal
spinal column, columna vertebral
spinal cord, espina dorsal, médula espinal
spinal nerve, nervio espinal
spindle, huso
spindle fiber, fibra del huso
spine, espinazo, espina, columna vertebral
spinneret, fileras
spiracle, espiráculo
spiral, espiral
spiral galaxies, galaxias espirales
spirochete, leptospiraceae
spirogyra, spirogyra
spit, espetar
spleen, bazo

splice, empalmar
split, dividir
spoiler, spoiler, frenos de aire
sponge, esponja
spongy bone, hueso esponjoso
spongy layer, capa esponjosa
spongy mesophyll, mesófilo esponjoso
spongy tissue, tejido esponjoso
spontaneous chemical change, cambio químico espontáneo
spontaneous combustion, autoinflamación
spontaneous generation theory, teoría de la generación espontánea
spontaneous ignition, ignición espontánea
spontaneous mutation, mutación espontánea
spontaneously, espontáneamente
spoon, cuchara
sporangium, esporangio
spore, espora
  spore reproduction, reproducción de esporas
sporogenesis, esporogénesis
sporophyte generation, generación de esporofitos
sporulation, esporulación
spot test for fat, ensayo de la mancha de grasa
sprain, esguince
spray, rociar
spread, propagar
spring scale, dinamómetro
spring tide, marea viva
spring[1], primavera
spring[2], resorte, muelle
spring[3], brincar, brotar
sprinkle, espolvorear
sprout, brote
spruce, picea
sputum, esputo
squall, ráfaga
square, cuadrado
  square root, raíz cuadrada
squid, calamar
squint, estrabismo, ser bizco, mirar de reojo, mirar entrecerrando los ojos
stability, estabilidad
  stability constant, constante de estabilidad
stabilization, estabilización
stabilizer, estabilizador
stable, estable, permanente, estacionario, caballeriza
  stable compound, compuesto estable
stack, pila
stages, etapas
stagnant, estancado
stain, mancha
staining, tinción
stainless steel, acero inoxidable
stake, estaca
stalactite, estalactita
stalagmites, estalagmitas
stalk[1], tallo (plant), rabo (fruit)
stalk[2], acechar
stamen, estambre
staminate, estaminífero
  staminate flower, flor estaminada

standard, estándar
standard atmospheric pressure, presión atmosférica normal
standard calomel electrode, electrodo normal de calomelanos
standard condition, condición estándar
standard condition for temperature and pressure (STP), condiciones normales de presión y temperatura (CNPT)
standard electrode potential, potencial normal de electrodo
standard heat of formation, entalpía estándar de formación
standard oxidation-reduction potential, potencial normal de oxidación-reducción
standard pressure, presión estándar
standard solution, solución estándar
standard time, hora estándar, hora oficial
standing[1], posición
standing[2], vertical, derecho
standing wave, onda estacionaria
stannic, estánico
stannous, estannoso
stapes, estribo
staphylococcus, estafilococo
star, estrella
star life cycle, ciclo de vida de una estrella
starch, almidón
starfish, estrella de mar
startling, alarmante
state, estado
state of matter, estado de agregación de la materia
statement, declaración
static, estático
static charge, carga estática
static electricity, electricidad estática
station model, estación de modelo
stationary, estacionario
stationary front, frente estacionario
stationary wave, onda estacionaria
statistics, estadística
statute mile, milla terrestre
STD (sexually transmitted disease), enfermedad de transmisión sexual (ETS)
steady state, estado de equilibrio
steam, vapor
steam distillation, destilación a vapor
steam engine, máquina de vapor
steam iron, plancha de vapor
steam shovel, excavadora
stearic acid, ácido esteárico
steel, acero
steel-engraving, grabado en acero
stellar, astral
stem, tallo

stem cell,
célula madre
step-down transformer,
transformador reductor
steppe, estepa
step-up transformer,
transformador elevador
stereochemistry, estereoquímica
stereoisomer, estereoisómero
stereomicroscope, microscopio estereoscópico
stereoscope, estereoscopio
sterile, estéril
sterility, esterilidad
sterilization, esterilización
sterilizing, esterilizante
sternum, esternón
steroid, esteroide
sterol, esterol
stethoscope, estetoscopio
stick, palo
stiffness, rigidez
stigma, estigma
still, todavía
stimulant, estimulante
stimulate, estimular
stimulated emission,
emisión estimulada
stimulus (pl.stimuli),
estímulo
stinger, aguijón
stinging cell,
célula urticante
stipule, estípula
stitch, suturar
stock, abastecer
stolon, estolón
stoma, estoma
stomach, estómago
stomata, estoma
stomate (stoma), estomas (estoma)
stone[1], piedra (rock)
Stone Age,
Edad de Piedra
stone[2], hueso (fruit)
stopper, obturador
stops, detiene
stopwatch, cronómetro
storage, almacenamiento
storage tissue,
tejido de almacenamiento
store, almacenar
storm surge, mareas de tempestad
STP (standard condition for temperature and pressure), CNPT (condiciones normales de presión y temperatura)
straight, recto
straight angle,
ángulo llano
straight-chain compound, cadena lineal compuesta
straighten, enderezar
strain, presión, tensión, carga, esguince, torcedura
strait, estrecho
strand, hebra
strata (sing. stratus),
estratos
strategy, estrategia
stratification,
estratificación
stratified, estratificada
strato, estrato
stratocumulus,
estratocúmulus
stratosphere, estratósfera
stratum (pl. strata),
estrato

stratus, estrato, stratus
stratus cloud, nube estrato
straw, paja
streak, veta
stream, corriente
stream bed, lecho del río
stream discharge, corriente de descarga
stream draining pattern, patrón de flujo de drenaje
stream flow, gasto, caudal
streamline, dar líneas aerodinámicas
strength, fuerza
strep throat, amigdalitis
streptococcus, estreptococo
streptomycin, estreptomicina
stress, presión, estrés
stretch, estiramiento
striated muscle, músculo estriado
striation, estriación
strike, golpe, descubrimiento
string, cuerda
string theory, teoría de cuerdas
strip, faja
strip cropping, cultivo en fajas
strip mine, mina a cielo abierto
strobe, estroboscópico
strobe light, luz estroboscópica
strobe photography, fotografía estroboscópica
stroke, derrame cerebral, apoplejía
stroma, estroma
strong acid, ácido fuerte
strong base, base sólida
strong electrolyte, electrolito fuerte
strong nuclear force, fuerza nuclear fuerte
strontium, estroncio
structural adaptation, adaptación estructural
structural formula, fórmula estructural
structure, estructura
strychnine, estricnina
sturgeon, esturión
style, estilo
subatomic, subatómico
subatomic particle, partícula subatómica
subclass, subclase
subcutaneous, subcutáneo
convergent boundary, borde convergente, límite convergente
subduction zone, zona de subducción
subkingdom, subreino
sublevel, subnivel
sublimate, sublimar
sublimation, sublimación
sublime, sublime
submarine, submarino
submarine canyons, cañones submarinos
submerge, sumergir
submergence, sumersión
suborder, orden subordinado
subphylum, subfilo
subset, subconjunto
subshell, semicapa
subsidence, hundimiento
subsoil, subsuelo
subspecies, subespecie

substance, sustancia
substitute, sustituto
substitution reaction, reacción de sustitución
substrate, sustrato
subtraction, resta, substracción
subtractive, sustractivo
subtrahend, sustraendo
subtropical, subtropical
subvert, invertir
succession, sucesión
succulent plant, planta carnosa, planta suculenta
sucker (zoology), psila
sucrase, sacarasa
sucrose, sacarosa
suction, succión
  suction pressure, presión de succión
suddenly, repentinamente
sufficiently, suficientemente
sugar, azúcar
sulfa drug, droga sulfa
sulfate, sulfato
sulfide, sulfuro
sulfur, azufre
  sulfur dioxide, dióxido de azufre
sulfuric acid, ácido sulfúrico
sum, suma
summarize, resumir
summer, verano
sun, sol
sundial, reloj de sol
sunny, soleado
sunspot, mancha solar
super continent, supercontinente
super gravity, supergravedad
super symmetry, supersimetría
superconductor, superconductor
supercooling, superenfriamiento
supercooled water, agua superenfriada
supergiant, supergigante
superior vena cava, vena cava superior
supernova, supernova
superoxides, superóxidos
supersaturated solution, solución sobresaturada
supersaturation, sobresaturación
supersonic, supersónico
supplementary angles, ángulos suplementarios
supplementation, suplementación, complemento
support[1], apoyo, respaldo
support[2], apoyar
surf, explorar, navegar
surface[1], superficie
  surface active agent, agente tensioactivo
  surface activity, actividad de la superficie
  surface area, área de la superficie
  surface chemistry, superficie química
  surface concentration excess, exceso de concentración en la superficie
  surface currents, corrientes superficiales
  surface orientation, orientación de la superficie

surface reaction, superficie de reacción
surface tension, tensión superficial
surface wave, onda de superficie
surface[2], emerger
surfactant, tensoactivo
surgery, cirugía
surplus, superávit
surrogate parent, padre sustituto
surround, rodear
surroundings, alrededores
survey, encuesta
survival, supervivencia
survival of the fittest, supervivencia del más apto
survive, sobrevivir, supervivencia
surviving, sobrevivir
suspect, sospechar
suspend, suspender
suspension, suspensión
sustainable use, uso sustentable
suture, sutura
swallow[1], trago
swallow[2], golondrina
swallow[3], tragar
swamp[1], ciénaga, pantano
swamp[2], agobiar
swamp[3], hundir
S-wave, onda S
sweat[1], sudor
sweat gland, glándula sudorípara
sweat[2], transpiración
sweat[3], transpirar
Swedish, sueco
sweep, barrer
swim-bladder, vejiga natatoria
swing[1], oscilar, balancearse
swing seat, asiento abatible
swing[2], oscilación
swirl, remolino
switch[1], cambio
switch[2], cambiar, intercambiar
switch[3], interruptor
symbiosis, simbiosis
symbiotic, simbiótico
symbol, símbolo
symmetrical, simétrico
symmetry, simetría
sympathetic, simpático
sympathetic nervous system, sistema nervioso simpático
symptom, síntoma
synapse, sinapsis
synchrocyclotron, sincrociclotrón
synchrotron, sincrotón
syncline, sinclinal
syndet, Syndet
syndrome, síndrome
synodic month, mes sinódico
synonym, sinónimo
synoptic weather map, mapa meteorológico sinóptico
synthesis, síntesis
synthesis gas, gas de síntesis
synthesize, sintetizar
synthetic, sintético
synthetic resin, resina sintética
syphilis, sífilis
syringe, jeringa

system, sistema
systematic, sistemático
    systematic method, método sistemático
systemic circulation, circulación sistémica
systole, sístole
systolic pressure, presión sistólica

T cell, célula T
table, mesa
    table salt, sal de mesa
tactile, táctil
tadpole, renacuajo
taiga, taiga
tail, cola
    tail fin, aleta
talc, talco
talon, garra
talus, astrágalo, talus
tangent, tangente
tank, tanque
tanker, camión cisterna
tantalum, tántalo
tap root, raíz principal
tape measure, cinta métrica
tapeworm, tenia, solitaria
tar, alquitrán, brea
    tar pit, fosos de alquitrán
tarantula, tarántula
tarsal, tarso
tartaric acid, ácido tartárico
taste bud, papila gustativa
Taurus, Tauro
taxonomy, taxonomía
Tay-Sachs disease, enfermedad de Tay-Sachs
tear[1], lágrima
tear[2], rasgar, desgarrarse
teat, teta
technetium, tecnecio
technique, técnica
technologist, tecnólogo
technology, tecnología
tectonic boundary, contacto tectónico
teeth (sing. tooth), dientes (sing. diente)
tektite, tectita
telecommunication, telecomunicación
telegraph, telégrafo
telemetry, telemetría
telescope, telescopio
telluride, telururo
tellurium, telurio
telophase, telofase
temperate, templado
    temperate deciduous forest, bosque templado caducifolio
    temperate zone, zona templada
temperature, temperatura
    temperature inversion, inversión de temperatura
    temperature scale, escala de temperatura
template, plantilla
temporal lobe, lóbulo temporal
temporary, temporalmente
    temporary magnet, imán temporal, imán temporario
tend, tender
    tend to, tender a

tendency, tendencia
tendon, tendón
tendril, zarcillo
tensile, tracción
tension, tensión
tentacle, tentáculo
terabyte, terabyte
terbium, terbio
terephthalic acid, ácido tereftálico
term, término
terminal, terminal, polo
  terminal bud, yema terminal
  terminal moraine, morrena terminal
  terminal velocity, velocidad terminal
termite, termes
ternary, ternario
  ternary acid, ácido ternario
  ternary compound, compuesto ternario
terrace, terraza
terraced, bancales
terrae, tierras
terrestrial, terrestre
  terrestrial biome, biomas terrestres
  terrestrial motions, movimientos terrestres
  terrestrial planets, planetas terrestres
  terrestrial radiation, radiación terrestre
terrigenous, terrígeno
territory, territorio
tertiary, terciario
  tertiary alcohol, alcohol terciario
Terylene, Terylene
Tesla, Tesla
test, test, prueba, examen
  test cross, retrocruzamiento
  test tube, tubo de ensayo
  test tube holder, soporte del tubo de prueba
  test tube rack, gradilla
testicle, testículo
testis, testículo
testosterone, testosterona
tetanus, tétanos
tetrachloride, tetracloruro
tetrad, tétrada
tetraflouroethylene, tetrafluoroetileno
tetragonal, tetrágono
  tetragonal system, sistema tetragonal
tetrahedron, tetraedro
tetramer, tetrámero
tetraploid, tetraploide
texture, textura
thalamus, tálamo
thalassemia, talasemia
thallium, talio
theorem, teorema
theoretical physicist, físico teórico
theoretical plate, plato teórico
theory, teoría
  theory of relativity, teoría de la relatividad
  theory of use and disuse, teoría del uso y desuso
therapy, terapia
thermal, térmico
  thermal conductivity, conductividad

calórica
thermal diffusion, difusión térmica
thermal energy, energía térmica
thermal equilibrium, equilibrio térmico
thermal expansion, expansión térmica
thermal pollution, contaminación térmica
thermal polymerization, polimerización térmica
thermal vent, fuente hidrotermal
thermion, termión
thermite, termita
thermobalance, termobalanza
thermochemistry, termoquímica
thermocouple, termocupla
thermodynamics, termodinámica
thermoelectric, termoeléctrico
thermoform, termoformado
thermograph, termógrafo
thermometer, termómetro
thermonuclear, termonuclear
thermonuclear device, dispositivo termonuclear
thermonuclear reaction, reacción termonuclear
thermoplastic, termoplástico
thermosetting, termoendurecibles
thermosphere, termósfera
thermostat, termostato
thiamine, tiamina
thicker, más grueso
thickness, grosor
thigh, muslo
thin, delgado
thin film, película delgada
thiol, tiol
third-level consumer, consumidor de tercer nivel
thistle tube, tubo de cardo
thoracic duct, conducto torácico
thorax, tórax
thorium, torio
thorn, espina
thread, hilo
threshold, umbral, límite
throat, garganta
thrombin, trombina
thromboplastin, tromboplastina
thrombosis, trombosis
thrombus, trombo
thrust, empuje
thulium, tulio
thumb, pulgar
thunder, trueno
thunderstorm, tormenta eléctrica
thymine, timina
thymus, timo
thymus gland, timo
thyroid, tiroides
thyroid-stimulating hormone (TSH), hormona estimulante de la tiroides
thyroxine, tiroxina
tibia, tibia
tick, garrapata

tidal energy, energía mareomotriz
tidal pool, poza de marea
tidal wave, maremoto, marejada
tide, marea, corriente
tighten, apretado, ajustado
tightly, estrechamente
till, hasta
tilt[1], inclinar, volcar
tilt[2], inclinación, vuelco
tilted strata, estratos inclinados
timberline, límite del arbolado
timbre, timbre
time, tiempo, período de tiempo, momento, época
 time zone, zona horaria
tin, estaño
tincture, tintura
tiny, diminuto
tipping point, punto crítico, punto límite
tire, neumático
tissue, tejido
 tissue culture, cultivo de tejidos
 tissue fluid, fluido de tejidos
titanium, titanio
titrant, valorante
titration, valoración
TNT, trinitrotolueno, TNT
toad, sapo
tobacco, tabaco
toe, dedo del pie
toenail, uña del pie
tolerate, tolerar
toluene, tolueno
tombolo, tómbolo
ton, tonelada
toner, tóner
tongs, tenazas
tongue, lengua
 tongue rolling, enrollar la lengua
tonsil, tonsila, amígdala
tonsilitis, amigdalitis
tool, herramienta
tooth (pl. teeth), diente (pl. dientes)
 tooth decay, caries
 tooth root, raíz del diente
toothpick, palillo
topaz, topacio
topography, topografía
topology, topología
topsoil, suelo
tornado, tornado
torque, par motor
torr, milímetro de mercurio
torsion, torsión
 torsion balance, balanza de torsión
tortoise, tortuga
torus, toro (matemáticas)
total internal reflection, reflexión total interna
touch[1], toque, tacto
touch[2], tocar, afectar
tower, torre
toxic, tóxico
toxicology, toxicología
toxin, toxina
trace element, elemento traza
tracer, trazador
trachea, tráquea
tracheophyte, traqueófita, planta vascular
track, pista, vía, rastro

tract, tracto
tractor, tractor
trade winds, vientos alisios
trade-off, compensación
traffic, tráfico
train, tren
trait, rasgo, característica
trajectory, trayectoria
trampoline, trampolín
tranquilizer, tranquilizante
transcription, transcripción
transducer, transductor
transduction, transducción
transfer[1], transferir
transfer[2], transferencia
  transfer pipette, pipeta de transferencia
  transfer RNA (tRNA), ARN de transferencia (tARN)
transference number, número de transferencia
transform, transformar
  transform boundaries, transformación de límites
transformation, transformación
transformer, transformador
transfusion, transfusión
transgenic, transgénico
transistor, transistor
transition, transición
  transition element, elemento de transición
  transition series, serie de transición
  transition zone, zona de transición
translocation, translocación
translucent, translúcido
transmission, transmisión
transmit, transmitir
transmitted wave, onda transmitida
transmitter, emisora, transmisor
transmutation, transmutación
transparent, transparente
transpiration, transpiración
  transpiration pull, jalado del agua por transpiración
transplant, transplante
transport[1], transportar
transport[2], transporte
  transport system, sistema de transporte
transported sediment, sedimento transportado
transported soil, suelo transportado
transporting system, sistema de transporte
transuranic element, elementos transuránicos
transverse, transverso
  transverse colon, colon transverso
  transverse section, sección transversal
  transverse wave, onda transversal
trap number, trap en electrónica
trapezium, trapecio
trapezoid, trapezoide
trauma, trauma
treat, tratar
treatment, tratamiento
tree, árbol
trellis pattern, patrón de enrejado

tremor, temblor
trench[1], atrincherar
trench[2], trinchera, zanja
trend, tendencia, dirección, curso
triad, tríada
trial, ensayo, prueba
triangle, triángulo
triangulation, triangulación
tributary, afluente, tributario
triceps, tríceps
triceratops, triceratops
trichinosis, triquinelosis, triquinosis
trichloroethylene, tricloroetileno
trichloroflouromethane, trichloroflouromethane
triclinic, triclínico
tricresyl phosphate, fosfato tricresilo
tricuspid, tricúspide
tricyclic, tricíclicos
triethanolamine, trietanolamina
triethylaluminum, triethylaluminum
trigger, gatillo, disparador
triglyceride, triglicérido
trigonometric function, función trigonométrica
trigonometry, trigonometría
trihydroxy alcohol, alcohol trihidroxi
trillion, billón, trillón
trilobite, trilobite
trioxide, trióxido
triple beam balance, balanza de tres brazos
triple bond, enlace triple
triple point, punto triple
triplet code, código de tripletes
triploid, triploide
tritium, tritio
trivalent, trivalente
trombone, trombón
trophic level, nivel trófico
tropic, trópico
  Tropic of Cancer, trópico de Cáncer
  Tropic of Capricorn, trópico de Capricornio
tropical, tropical
  tropical zone, zona tropical
tropism, tropismo
troposphere, tropósfera
trough, borrasca, zona de baja presión
true solution, solución verdadera
trumpet, trompeta
trunk (tree), tronco
trypanosome, trypanosomatida
trypsin, tripsina
tsunami, tsunami
tube, tubo
tuber, tubérculo
tuberculosis, tuberculosis
tubing, tubería
tubule, túbulo
tuff, tufo
tumor, tumor
tundra, tundra
tune, melodía
tungsten, tungsteno
  tungsten carbide, carburo de tungsteno
tunicate, tunicado
tuning fork, diapasón
tunnel, túnel
turbidites, turbiditas
turbidity, turbidez
  turbidity currents,

corrientes de turbidez
turbine, turbina
turbojet, turborreactor
turbulent, turbulento
turbulent flow, flujo turbulento
turf, césped
turgid, turgente
turgor pressure, presión de turgencia
Turner's syndrome, síndrome Turner
turpentine, trementina
turtle, tortuga
tusk, colmillo
twice, dos veces
twig, ramita
twirl, giro
twist, torcedura
two-hole stopper, tapón de dos orificios
type, tipo
typhoid, tifoidea
typhoid fever, fiebre tifoidea
typhoon, tifón
typhus, tifus
tyrannosaurus, tiranosaurio
tyrosine, tirosina

udder, ubre
ulcer, úlcera
ulna, cúbito
ultimate analysis, última instancia
ultracentrifuge, ultracentrífuga
ultrafiltration, ultrafiltración
ultrasonography, ecografía
ultrasound, ultrasonido
ultraviolet, ultravioleta
umbel, umbela
umbilical cord, cordón umbilical
umbilicus, ombligo
umbra, umbra
unaffected, inafectado
unavailability, sin disponibilidad
unbalance, desequilibrar
unbalanced forces, fuerzas desiguales
uncertainty, incertidumbre
uncertainty principle, principio de incertidumbre
unchanged, sin cambio
unconformity, inconformidad
undefined target, meta no definida
under, debajo
undergo, someterse a
undergrowth, maleza
undertow, resaca
undiminished, no disminuido
unequal, desigual
unglazed porcelain plate, plato de porcelana sin esmaltar
ungulate, ungulado
unicellular, unicelular
unified field theory, teoría del campo unificado
uniform, uniforme
uniform acceleration, aceleración uniforme
uniform circular motion, movimiento circular uniforme

uniform dispersion, dispersión uniforme
uniformly, uniformemente
union, unión, sindicato
unique, único
unit, unidad de fuerza
units of force, unidades de fuerza
univalent, univalente
univalve, univalvo
universal recipient, receptor universal
universal time, horario universal
universe, universo
unknown, desconocido
unsaturated, no saturado
unsaturated compound, compuesto insaturado
unsaturated fat, grasa no saturada
unsaturation, insaturación
unsorted, sin clasificar
unstable, inestable
unstable compound, compuesto inestable
unsteady, inestable
uplifting force, fuerza de elevación
upright, vertical
upward, arriba
uracil, uracilo
uranium, uranio
Uranus, Urano
urban desert, desierto urbano
urbanization, urbanización
urea, urea
urease, ureasa
ureter, uretra
urethra, uretra
uric acid, ácido úrico
urinary bladder, vejiga
urinary system, sistema urinario
urinary tract, vías urinarias
urine, orina
Ursa Major, Osa Mayor
usage, uso
use, uso
U-shaped valley, valle en forma de U
utensil, utensilio
uterine lining, revestimiento del útero
uterus, útero
utilize, utilizar
UV index, índice UV
uvula, úvula

vacancy, vacante
vacant, vacante
vaccinated, vacunado
vaccination, vacunación
vaccine, vacuna
vacuole, vacuola
vacuum, vacío
vacuum bottle, termo
vacuum cleaner, aspiradora
vacuum condensing point, punto de condensación al vacío
vacuum crystallization, cristalización al vacío
vacuum crystallizer, cristalizador de vacío
vacuum distillation, destilación al vacío

vacuum tube, tubo de vacío
vagina, vagina
vagus nerve, nervio vago
valence, valencia
valid, válida
valine, valina
valley, valle
valley glacier, valle glaciar
valuable, valiosa
value, valor
valve, válvula
Van Allen belt, cinturón de Van Allen
Van der Waals force, fuerza de Van der Waals
vanadium, vanadio
vane, veleta
van't Hoff equation, ecuación de van't Hoff
van't Hoff isochore, isocoro de van't Hoff
van't Hoff isotherm, isoterma de van't Hoff
vapor, vapor
vapor pressure, presión de vapor, presión de saturación
vapor pressure depression, vapor de presión negativa
vapor state, estado de vapor
vaporization, vaporización
vaporize, vaporizar
variability, variabilidad
variable, variable
variable factor, factor variable
variable star, estrella variable
variation, variación
variegated leaf, hoja abigarrada
variety, variedad
various, diferentes
vas deferens, conducto deferente
vascular, vascular
vascular bundle, haz vascular
vascular cambium, cámbium vascular
vascular cylinder, cilindro vascular
vascular plant, planta vascular, traqueófita
vascular ray, rayo vascular
vascular system, sistema vascular
vascular tissue, tejido vascular
vasoconstriction, vasoconstricción
vasodilation, vasodilatación
vasopressin, vasopresina
vector, vector
vector quantity, cantidad vectorial
vector resolution, resolución vectorial
vector sum, suma vectorial
vectorial, vector
vegetable, vegetal
vegetable kingdom, reino vegetal
vegetative propagation, multiplicación vegetativa
vehicle, vehículo
vein, vena

veldt, meseta de escasa pluviosidad de Sudáfrica
velocity, velocidad
velocity of light, velocidad de la luz
vena, vena
vena cava, vena cava
venation, venación
venereal disease, enfermedad venérea
venom, veneno
venous flow, flujo venoso
ventral, ventral
ventral blood vessel, vaso sanguíneo ventral
ventral nerve cord, cordón nervioso ventral
ventral root, raíz ventral
ventricle, ventrículo
venule, vénula
Venus, Venus
Venus flytrap, Venus atrapamoscas
verification, verificación
vernal equinox, equinoccio de primavera
vertebra, vértebra
vertebral column, columna vertebral
vertebrate, vertebrado
vertex (pl. vertices), vértice
vertical angles, ángulos verticales
vertical rays, rayos verticales
vertically, verticalmente
vesicle, vesícula
vessel, vaso
vestiges, vestigios
vestigial, vestigial, rudimentario
veterinary medicine, medicina veterinaria
vibrate, vibrar
vibration, vibración
vibrational, vibracional
vibrational motion, movimiento vibratorio
vicinal, vecinal
vigorously, vigorosamente
villi, vellosidad
villus, vellosidades
vinegar, vinagre
vinyl, vinilo
vinyl chloride, cloruro de vinilo
violate, contravenir, violar
viral disease, enfermedad viral
Virgo, Virgo
virtual, virtual
virtual image, imagen virtual
virus, virus
viscera, víscera
visceral muscle, músculo visceral
viscosity, viscosidad
viscous, viscoso
viscous liquid, líquido viscoso
visible, visible
visible light, luz visible
visible spectrum, espectro visible
visualize, visualizar
vital, vital
vital capacity, capacidad vital
vitamin, vitamina
vitreous, vítreo
vitreous humor, humor vítreo

viviparous, vivíparo
vivisection, vivisección
vocal, vocal
  vocal cords, cuerdas vocales
voice box, caja de voz (laringe)
volatile, volátil
  volatile liquid, líquido volátil
volatilize, volatilizar
volcanic arc, arco insular
volcanic ash, ceniza volcánica
volcanic neck, cuello volcánico
volcano, volcán
volt, voltio
voltage, voltaje
voltaic, voltaico
  voltaic cell, celda galvánica
voltmeter, voltímetro
volume, volumen
  volume bottle, volumen de una botella
volumetric, volumétrico
  volumetric analysis, análisis volumétrico
  volumetric flask, matraz aforado
  volumetric pipette, pipeta volumétrica
voluntary, voluntario
  voluntary action, acción voluntaria
  voluntary behavior, comportamiento voluntario
  voluntary muscle, músculo voluntario
volunteer, voluntario
vortex, vórtice
vowel, vocal
V-shaped valley, valle en forma de V
vulcanization, vulcanización
vulcanize, vulcanizar
vulva, vulva

waist, cintura
wait, esperar
walking the outcrop, caminar por afloramientos
walnut, nuez, nogal
waning, menguante
warm, tibio
  warm blooded, de sangre caliente
  warm front, frente cálido, frente caliente
warning coloration, coloración de advertencia
warren, madriguera
wart, verruga
wasp, avispa
waste, residuo
water, agua
  water budget, balance hídrico
  water cycle, ciclo del agua
  water flea, pulga de agua
  water gas, gas de agua
  water glass, vaso de agua
  water of crystallization, agua de cristalización

water of hydration, agua de hidratación
water pollution, contaminación hídrica
water potential, potencial hídrico
waterproof, impermeable
water purification, purificación del agua
water shed, cuenca de drenaje
water softening, ablandamiento del agua
water table, capa freática
water vapor, vapor de agua
water vascular system, agua del sistema vascular
waterfall, cascada
watershed, línea divisoria de aguas
waterspout, tromba marina
watt, vatio
watt-hour, vatio-hora
wattage, potencia en vatios, vataje
wave, ola
wave mechanics, mecánica ondulatoria
wave refraction, refracción de ondas
wave speed, velocidad de las olas
wave velocity, velocidad de la onda
wavelength, longitud de onda
wax, cera
wax layer, capa de cera
waxing, encerar, depilar
weak acid, ácido débil
weak base, base débil
weak electrolyte, electrolito débil
weak force, fuerza débil
weak nuclear force, fuerza nuclear débil
weasel, comadreja
weather, clima
weather forecasting, pronóstico del tiempo
weather vane, veleta
weathering, desgaste
weathering agents, agentes atmosféricos
web[1] (computer), Web
web[2] (spider), tela
webpage, página web
website, sitio web
wedge, cuña
weed, maleza, yuyo
weed killer, herbicida
week, semana
weighing bottle, botella de pesaje
weight, peso
Western Hemisphere, hemisferio occidental
wet cell, pila líquida
wet-bulb depression, depresión de bulbo húmedo
wetlands, tierras de pantanos
wet-mount slide, muestra en montaje húmedo
whale, ballena
whalebone, barba de ballena
wheat, trigo
wheel and axle, rueda y eje

whirlpool, remolino, vórtice
white, blanco
white blood cell, leucocito
white corpuscle, glóbulo blanco
white dwarf, enana blanca
white matter, sustancia blanca
whole blood, sangre entera
whole number, número entero
whooping cough, tos ferina
widen, ampliar, ensanchar
width, ancho
wildlife conservation, conservación de la vida silvestre
wilt, marchitarse
wind, viento
wind break, rompevientos
wind chill factor, factor frío del viento
wind erosion, erosión eólica
wind pollination, polinización anemófila, polinización por el viento
wind tunnel, túnel de viento
wind turbines, turbinas de viento
wind vane, veleta
wind-chill factor, factor de enfriamiento del viento
windmill, molino de viento
window-sill, antepecho de ventana
windpipe, tráquea
windward, viento
wing, ala
wingspan, envergadura
winter, invierno
winter solstice, solsticio de invierno
wire gauze, alambre de gasa
wisdom tooth, muela del juicio
wise, sabio
wishbone, espoleta
within, dentro de
wolfram, wolframio, tungsteno
womb, matriz, útero
wood, madera, bosque
wood alcohol, alcohol de madera (metanol)
woody fiber, fibra leñosa
woody stem, tallo leñoso
woolly mammoth, mamut lanudo
work[1], trabajo
work[2], trabajar
worker ant, hormiga obrera
worker bee, abeja obrera
World Wide Web, World Wide Web
worm, gusano, lombriz
wrinkle, arruga
wrist, muñeca

xanthophyll, xantófila
xanthoproteic test, prueba xantoproteica
x-axis, eje x
x-chromosome, cromosoma X
xenon, xenón
Xerox, fotocopiar

x-ray[1], rayos X
x-ray[2], radiografiar
X-value, valor X
xylem, xilema
xylene, xileno

yard (measurement), yarda
y-axis, eje y
y-chromosome, cromosoma Y
year, año
yeast, levadura
yellow, amarillo
yellow fever, fiebre amarilla
yew, tejo
yolk, yema
  yolk sac, saco vitelino
young landscape, paisaje joven
youth, juventud
ytterbium, iterbio
yttrium, itrio
Y-value, el valor Y

z-axis, eje z
zebra mussel, mejillón cebra
zenith, cénit
zeolite, zeolita
zero, cero
  zero gravity, gravedad cero, gravedad ingravidez
  zero group, grupo cero, grupo de la tabla periódica
zinc (Zn), cinc (Zn)
  zinc oxide, óxido de cinc
zincate, cincato
zircon, circón
zirconium, circonio
zodiac, zodiaco
zonation, zonificación
zone, zona
  zone of aeration, zona de aireación
  zone of convergence, zona de convergencia
  zone of divergence, zona de divergencia
  zone of saturation, zona de saturación
zoo, zoológico
zoology, zoología
zooplankton, zooplancton
zygote, cigoto

Velázquez Press

For over 150 years, *Velázquez Spanish and English Dictionary* has been recognized throughout the world as the preeminent authority in Spanish and English dictionaries. Velázquez Press is committed to developing new bilingual dictionaries and glossaries for children, students, and adults based on the tradition of *Velázquez Spanish and English Dictionary.*

We invite you to go to www.AskVelazquez.com to access online services such as our free translator and member forum. We also invite you to add on to the glossary. If you know of a K-12 science term that is not included, please let us know at info@academiclearningcompany.com for future editions.

Durante más de 150 años, *Velázquez Spanish and English Dictionary* ha sido reconocido como la máxima autoridad en diccionarios de español e inglés en todo el mundo. Velázquez Press está comprometido a elaborar diccionarios y glosarios bilingües para niños, estudiantes y adultos en la tradición del *Velázquez Spanish and English Dictionary.*

Lo invitamos a visitar www.AskVelazquez.com para acceder a los servicios en línea como nuestro traductor automático y el foro. Si sabe de algún término de ciencia de los grados escolares kinder a 12 que no está incluido, por favor, mande un correo electrónico a info@academiclearningcompany.com para ediciones futuras.

Other Velázquez Resources

Velázquez Spanish and English Dictionary

- More than 250,000 entries with accessible pronunciation guides for BOTH Spanish and English
- Revised for the 21st century
- Thumb-indexed to help user find words faster
- Covers regional variations of US, Latin American and European Spanish
- For teachers and students, notes on grammar integrated into the main body
- Entries include examples for better understanding of connotations and usage

BOOK INFORMATION

- **ISBN 13: 978-1-5949-5000-1 • Pub Date: August 2007**
- **Format: Hardcover, 2,008 pages • Size: 7.5 in. x 9.5 in.**
- **Price: $29.95 USD**

Velázquez Large Print Spanish and English Dictionary

- Carries the Seal of Approval of NAVH
- Over 38,000 easy-to-read bold entries with guide phrases
- Great for mature and visually impaired readers and students
- Includes pronunciation keys in BOTH the Spanish and
- English sections
- Easy-to-use format and up-to-date
- User's Guide included
- Durable hardcover binding for long-lasting use

BOOK INFORMATION

- **ISBN 13: 978-1-5949-5002-5 • Pub Date: January 2006**
- **Format: Hardcover, 1,006 pages • Size: 8.25 in. x 9.5 in.**
- **Price: $22.95 USD**

Velázquez World Wide Spanish and English Dictionary

- Word-to-word translation for state standardized testing
- Contains no offensive words. Adopted by many states for classroom use
- Over 85,000 colored entries for easy-to-find reference
- BOTH Spanish and English pronunciation keys and explanation of sounds

BOOK INFORMATION

- **ISBN 13: 978-1-5949-5001-8 • Pub Date: March 2010**
- **Format: Paperback, 691 pages • Size: 5.25 in. x 6.8 in.**
- **Price: $12.95 USD**

Velázquez Spanish and English Dictionary, Pocket Edition

- Over 75,000 entries and 110,000 translations in a convenient pocket size
- Includes BOTH Spanish and English pronunciation keys, and explanation of sounds
- Spanish and English grammar synopses to better understand words in context
- Great for students, office, or day-to-day use

BOOK INFORMATION

- **ISBN 13: 978-1-5949-5003-2 • Pub Date: May 2006**
- **Format: Paperback, 640 pages • Size: 4.25 in. x 7 in.**
- **Price: $5.98 USD**

Velázquez Spanish and English Glossary for the Social Studies Classroom

- This 15,000 entry reference tool is the only one of its kind on the market
- Word to word translations format to meet state standardized tests
- Translations of multiple-word terms
- Contains a broad range of social studies terms covering everything from U.S. history and geography to economics and psychology

BOOK INFORMATION

- **ISBN 13: 978-1-59495-014-8 • Pub. Date: May 2010**
- **Format: Paperback • Size: 5 in x 8 in**
- **Price: $12.95 USD**

Velázquez Spanish and English Glossary for the Mathematics Classroom

- Over 10,000 mathematical translations
- Word to word translations format to meet standards of state standardized tests
- Translations of multiple-word terms
- Extensive coverage of mathematic terms in the areas of arithmetic, logic, algebra, geometry, trigonometry, statistics, and calculus
- Covers words that apply to the mathematic and scientific method
- Includes common vocabulary for the mathematics classroom

BOOK INFORMATION

- **ISBN 13: 978-1-59495-017-9 • Pub Date: May 2010**
- **Format: Paperback, 274 pages • Size: 5 in. x 8 in.**
- **Price: $12.95 USD**

Velázquez Academic Vocabulary Sheet

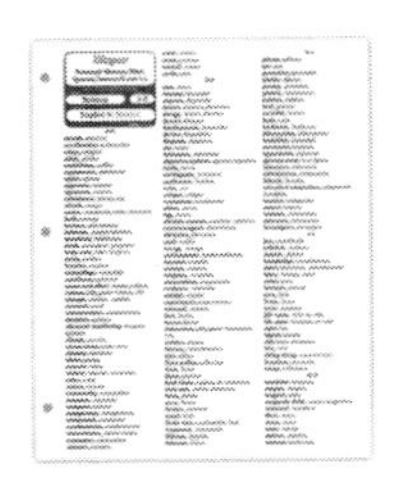

- 300 Key Bilingual Academic Vocabulary for ESL
- Used in state standardized tests and classroom learning
- 90 languages available in 3 different subjects (math, science, and language arts) and 3 grade levels (3-5, 6-8, and 9-12)
- 3 hole punched and laminated for Notebook Insert

BOOK INFORMATION

- **Pub Date: May 2011**
- **Format: laminated, 3 hole punched • Size: 8.5 in. x 11 in.**
- **Price: $4.98 USD**

Velázquez Spanish and English Glossary for the Language Arts/ESL Classroom

- Over 163,000 translations, including 5,000 key academic vocabulary terms for language arts
- Word-to-word translation format to meet standards of state standardized tests
- Translations of multiple-word terms
- Extensive coverage of language arts terms in the areas of English literature, psychology, philosophy, economics, and the arts

BOOK INFORMATION

- **ISBN 13: 978-1-59495-015-5 • Pub. Date: Fall 2011**
- **Format: Hardcover • Size: 5 in x 8 in**
- **Price: $16.95 USD**